About the Cover: Displayed on the cover is a sample of chonomic structures—the three observed neutrinos—electron neutrino, muon neutrino, and tauon neutrino. Chonomic structures are very powerful—they present the yachon-echon structure of the particle, the energy state of each sub-particle, the intrinsic spin of each sub-particle, each orbital spin of the particles, and the calculated and measured masses of the particle. As listed in the book, they also include the quark predictions for the particle.

Predicting the Masses
Volume 1
Introducing Chonomics

By Gordon L. Ziegler and Iris Irene Koch

Predicting the Masses
Volume 1
Introducing Chonomics

©2015 Gordon L. Ziegler and Iris Irene Koch

Last revised February 20, 2015

Authors

Gordon L. Ziegler and Iris Irene Koch
P.O. Box 1162
Olympia, WA 98507-1162 USA
e-mail: ben_ent100@msn.com

PREFACE

The masses of every known particle (as of 2008) are arranged to be calculated from first principles in this book set (*Predicting the Masses*, Volume 1, Introducing Chonomics; Volume 2, Predicting the Mesons; and Volume 3, Predicting the Baryons) making use of the data in "Prediction of the Masses of Every Particle, Step 1", by the authors and *Advanced Electrino Physics* Draft 3. This feat is impossible in the Quark Model, the Standard Model, the String Theory, and the Many Dimensional Theory. It is possible only with the Electrino Fusion Model of Elementary Particles with the Electrino Hypothesis that fracton charges come in $\pm$ e, $\pm$ e/2, $\pm$ e/4, and $\pm$ e/8, not the Quark Hypothesis that fracton charges come in $\pm$ 2e/3 and $\pm$ e/3.

All particle bonds in this book set (except for pions) are assumed by the authors to be orbital bonds. The strong nuclear force is active in nuclei (entities with two or more particles with dot unitons in them, but not in composite particles with only zero or one particles with a dot uniton in them. The strong nuclear force derives from the strong gravitational force and is mediated by the pion. The strong nuclear force is a stick-on force—like magnetism, but we do not have to be concerned with the strong nuclear force in this book set (*Predicting the* Masses, Volumes 1-3.) We do have to be concerned with magnetism a few times in this set—mostly not. Some calculational basics, however, remain the same in the new system: For the different particle types, there are at least three different spin relations, which are taken as postulates.

The errors in the calculation of masses may be due to the fact that the calculations are solely with the n and s parameters, omitting the l (elliptical) and m (tilt or magnetic) parameters and relativistic considerations. The calculations may be more accurate with those. All that is used in this book are the Electrino Hypothesis, the Niels Bohr style of mass calculations, the spin postulates, the measured mass of the electron, the different forces, and algebra. The results generally are calculated masses at one to

4

four place accuracies as compared to the measured masses of the particles, where measurements have been made.

Chapters 1 to 10 in this book are a road map on how to calculate the masses of particles. Of aid to the investigator are *Electrino Physics* Draft 2 by Gordon L. Ziegler and Iris Irene Koch and *Advanced Electrino Physics* Draft 3 by Gordon L. Ziegler and Iris Irene Koch. Chapters 7 and 8 in this book revise Appendix A and Appendix B in the first work, employing a tenth criteria in deriving the structures of particles—namely, forcing gravitons and other particles to not have any sub particles in ground state so the particles can be calculated and not needing to be defined. The first work lists the structures of the masses to be calculated in Appendix B (updated by the work in Chapter 8 of this book). The second work, *Advanced Electrino Physics*, Draft 3, solves for the masses of 22 particles including the pion through orbital bonds only (except for the pion), and providing a bank of g/2 factor calculations in Chapter 5 useful in calculating the masses of particles.

Particle physics is exploding with seemingly endless particle decay modes for massive particles, with no known periodic chart of the elementary particles to guide the particle physicists. This book set will supply the lack and make sense of particle physics for the theorist and the experimenter.

Chapter 16 applies the road map for calculating the masses of particles from first principles to neutrinos and anti-neutrinos. We find that neutrinos are minus imaginary masses and anti-neutrinos are positive imaginary masses.

Chapter 17 solves for three very different structures of gravitons, and discovers that all three different types of gravitons have the same array of particle masses and the same minimum mass—about 70.0 MeV per graviton. This is not insignificant or negligible. The gravitons are all massive—on the order of lower mass mesons. The sea of gravitons in a galaxy would entail a significant mass to the galaxy that, while invisible and not luminous to us, nevertheless would add a significant galaxy mass to astronomers' galaxy orbital calculations.

CONTENTS

Chapter
Page

Chapter 1

Introducing Chonomics

Electrinos are postulated particles (not yet detected) that lie at the foundation of everything we can see, hear, taste, smell, or feel. The use of the word electrino is not intended to validate the extant use of the word to describe the assumed easy polarizability and ionizability of electron fragments in liquid hydrogen. Our model of electrinos agrees that they are tiny pieces of electric particles, but that they are not easily ionizable pieces of electrons. Our model (the electrino fusion model) has them in speed of light containment barriers, not possible of being ionized or blasted apart. Our apologies go to the original coiner of the name electrino. We shamelessly appropriate it to our model as the only descriptive and succinct word to describe the most important particles in the universe.

All matter, light, and gravitons are made up of electrinos. Electrinos are all tiny spheres of charge—but not solid spheres. They are all thin film spheres of charge like soap bubbles with calculable radii, charge, and mass (zero mass at rest and fractions of the minus imaginary Planck's mass at the speed of light relative to the aether. In our terms, the mass of the largest electrino (the uniton), derived from first principles in [1](pp. 181-185) is

$$M_0 \approx -i\, 2.176\ 51(13) \times 10^{-08}\ \text{kg.} \tag{1-1}$$

Its radius (derived from first principles in [1](pp. 181-185)) is

$$R_0 = 2GM_q/\text{-}c^2 = i(\hbar G/c^3)^{1/2} \approx i\, 1.616\ 199(97) \times 10^{-35}\ \text{m,} \tag{1-2}$$

where M_q is the strong mass of the particle due to charge alone (half of the mass of the particle—the other half of the mass of the

7

particle is from the kinetic motion relative to the aether), G is Newton's gravitational constant, and ħ is Planck's constant $h/2\pi$. M_0 is our term for the minus imaginary Planck mass of the whole particle, the uniton; R_0 is our term for the corresponding positive imaginary radius of the spherical thin film of charge in the uniton.

Semions, quartons, and octons (½, ¼, and ⅛ charges) are also of interest to us. From equations parallel to equations (6-7) through (6-10) in [1](p. 183) we see that, while the rest masses of all the electrinos relative to the aether are zero, the net imaginary total masses and radii of the particles at aether relative velocity c are:

$$m_{octon} = \text{⅛ } m_{uniton} \approx -i\ 2.720\ 637(02) \times 10^{-09}\ kg. \tag{1-3}$$

$$r_{octon} = \text{⅛ } r_{uniton} \approx i\ 2.020\ 2484(12) \times 10^{-36}\ m. \tag{1-4}$$

$$m_{quarton} = \text{¼ } m_{uniton} \approx -i\ 5.441\ 275(03) \times 10^{-09}\ kg. \tag{1-5}$$

$$r_{quarton} = \text{¼ } r_{uniton} \approx i\ 4.040\ 4975(24) \times 10^{-36}\ m. \tag{1-6}$$

$$m_{semion} = \text{½ } m_{uniton} \approx -i\ 1.088\ 25(07) \times 10^{-08}\ kg. \tag{1-7}$$

$$r_{semion} = \text{½ } r_{uniton} \approx i\ 8.080\ 99(49) \times 10^{-36}\ m. \tag{1-8}$$

All electrinos not only have zero rest masses, they also have zero spins. The only way they can contribute detectable spins to the system is if they are in orbit in the particle. Total masses and spins can be calculated higher than the detectable masses and spins, but typical spin reactions in particles model after detectable spins.

All detectable particles are mass singularities and are in one, up to three, concentric particle scale black holes. The speed of light barriers of the black holes put matter on a solid foundation. The particle black holes renormalize the masses and spins of particles, because the observer cannot see through a black hole to see a mass or spin on the opposite side of the black hole. He/she can only see on the front side of the black hole at or outside the event horizon of the black hole. For instance, the total spin of the electron (counting all sides of the black hole) is $\pm$ 1 ℏ, but the observable spin of the electron is ½ ℏ, which is the value utilized in all the particle reactions. Also the masses of all faster than the speed of light orbital velocities are within the black hole, and are not observable, but the overall binding orbit of the sub particles is of the slower than the speed of light velocities relative to the aether, where opposite charges attract, and where the masses and spins of the orbit are detectable.

Let us now define the terms that are used in the system of chonomics (structure, decay, and reactions of particles). The word chonomics is taken from the common roots of two different kinds of whole particles in the Hebrew. A unity of a whole made up of a sole, single, but one entity is called yachid in the Hebrew. We take the name of a whole particle made up of only one particle as a yachon. A unity made up of parts is called an echad. We take a particle whole made up of parts as an echon. [1](Chapter 6). Notice that both yachons and echons end in chons. Thus we take a particle study of whole particles as chonomics. While allowing for various sub particles in the particles, chonomics is the structure, decay, and reactions of whole particles. Fractional particles can only be in the speed of light barrier containments, except for octons, but whole particles are free to combine or disassociate as the need may arise, recombine to different whole particles.

The masses of light particles cannot be calculated in the Quark Model, The Standard Model, the String Theory and the Many Dimensional Theory. It is not that the physicists have not yet figured out how to do it in those models. It is because it is impossible in those models. But it is possible to do this in a new Theory of Particle Physics—The Electrino Fusion Model of Elementary Particles. That feat is done in a prior paper ("Prediction of the Masses of Every Particle, Step 1") and this book (*Predicting the Masses*) without tensors, matrices, Hamiltonians, Schrödinger's Equations, Isospin and many other advanced mathematical tools and concepts. This book will use only the Niels Bohr style of particle mass calculation, three spin postulates, the measured mass of the electron, algebra, and the

9

Electrino Hypothesis—that fractional charged particles come in ± e, ± e/2, ± e/4, and ± e/8, not in ± 2e/3 and ± e/3 of the Quark Hypothesis. There were errors in all previous works by the authors on this subject. The particle force bonds in this book will be orbital bonds only (except for pions). This book is the most up-to-date work by the authors so far (02/20/2015). But we will not warrant this book to be error-free either.

The next thing to consider is that every known particle (except photons) can be constructed with various states of electrons, various states of pions, various states of neutrons, and various combinations of those particles. Because in the new theory there is a postulate that smooth symmetrical charge distributions cannot have detectable spin, the theory does not allow electrons to be spinning point charges. In the new theory electrons are composed of two half particles (semions) orbiting about each other at the speed of light. Pions are composed of four fourth charges (quartons)—with more than one possible spin orientation. A neutron is composed of a whole e particle (uniton) orbited by an electron (which is composed of two half charges orbiting about each other). Now if we could predict the masses of various states of electrons, pions, and neutrons, and learn how to put them together in compound particles and learn how to calculate the masses of multi-particle particles, we would learn how to predict the masses of almost every particle (photons excepted). We did the first step in the "Step 1" paper. Through the Electrino Hypothesis, Neils Bohr style of mass calculations, and algebra, we predicted the known electron, pion, and neutron family members to two to four place accuracies. Our challenges in this book are 1) to make a road map on how to do all the step 2 calculations of all the known particles in the first six states (states 0, 1, 2, 3, 4, 5) and 2) to calculate the masses of all known particles as of 2008.

The quark and lepton model of particle physics divides charges in quarks to ± 2e/3 and ± e/3. The electrino model of particle physics does not do that. Instead, it divides charges in electrinos to ± e, ± e/2, ± e/4, and ± e/8. The electrino model of particle physics does not hold that the quark and lepton model of particle physics is correct. Nevertheless, to facilitate cross referencing with the existing data, this book will employ quark model titles and classifications in the subsequent classification of particles.

The chonomic structures contained in the following material are the author's, but most of the particle data come from [1]. The author's chonomic structures in this book are induced from the following ten criteria: particle charge, spin, parity, mass, spin feasibility, preceding particles (to avoid duplication), decay schemes, the Pauli Exclusion Principle, the b-state laws, and the providential ruling that all particles in the Universe except electrons should be calculable and predictable from first principles without definition. The use of isospin in the precursor data, instead of the simple charge, made the author's work difficult; so too the convention of listing any charge π as π, and decay products of baryons as N . . ., where N can stand for many different baryons. For accurate results, please change to precise reporting conventions. These results are highly valuable, and worth doing right.

This book proves that all known matter, light, and gravitons can better be constructed of electrinos rather than quarks and leptons. All particles may be formulated with yachons and echons, or with +'s, -'s, o's, and •'s. The key to understanding chonomics is in [2], Chapter 10 and this book, chapters 1-10. .

The symbols in the grid are combinations of -, +, o, and •. These are not charge symbols—they represent the spins and sizes of the relevant particles. The symbol – stands for spin – ½ ħ, or – ½ h/2π, of a particle in the particle system. The symbol + stands for a particle with + ½ ħ spin. The symbol o stands for a particle with net zero orbital spin like the pion. The symbol • stands for a zero spin nearly point charge. The •'s are found only in state 2 energy levels.

This book is for particles only, not nuclei. The strong nuclear force (the force between like charged •'s) will not be studied in this book set. Except for photons, particles will have at most one • in them.

---

[1] J. Beringer *et al* (Particle Data Group) PR **D86**, 010001 (2012) and 2013 partial update for the 2014 edition (URL: http://pdg.lbl.gov).

[2] Gordon L. Ziegler, *Electrino Physics* (http://benevolententerprises.org Book List; 11/23/2013, Xlibris LLC; *Electrino Physics*, Draft 2 is now available at Amazon.com.).

Chapter 2

Positional Chonomic Grids

The authors here introduce a new kind of pictorial diagram different than Fineman diagrams used extensively in physics:

```
3        |
        ___
2        |
        ___
1        |
        ___
0        |
```

In the middle of the diagram is a vertical line. All symbols positioned on the left side of the line in the chonomic grid represent negatively charged particles. All symbols positioned on the right side of the vertical line in the chonomic grid represent positively charged particles. The symbols utilized in the diagram are only -, +, o, and ●. These are not charge symbols. Their charge information is portrayed by which side of the vertical line they are placed on. The different symbols portray different kinds of whole particles combined in the overall particle portrayed. The different symbols represent different spin and size characteristics of the whole particles considered in the overall particle. The − particle represents a sub whole particle with − ½ ℏ intrinsic spin. The + particle represents a sub whole particle with + ½ ℏ intrinsic spin. The o particle represents a net 0 orbital spin particle. We choose the o symbol instead of the 0 symbol because the orbits at rest are circular, not elliptical. They are elliptical only when they are in motion, but their long axis is not along the x axis, but is perpendicular to the x axis. Neils Bohr's system of electron orbits in Hydrogen were the easy circular orbits—which were made better and more precise by Sommerfeld taking into account elliptical orbits. Not so with electrino orbits. Circular orbits are adequate in this science.

The dot yachon ● is, like the o echon, a net 0 spin particle, but it is not an orbital system. It is a 0 spin nearly point charge particle.

The horizontal lines in the chonomic grid separate different energy states that a particle may be in. (The numbers on the left side of the chonomic grid aid in specifying the state numbers.) The bottom state is the 0 state. Only electrons or positrons (- or + particles [- ½ ℏ or + ½ ℏ spin electrons or positrons]) singly may be in the zero state, but − and + particles may appear in any or all elevated states if there are none in the 0 state. The reason for that is the particle masses are calculated partly through a sort of an exponential polynomial $b/n\alpha^{n/b}$, where b is the state number, and n is defined to correspond to b by the following arrangement:

b   0   1   2   3   4   5   6 . . .

n   0   1   3   6   10 15 21 . . .

j   0   1   2   3   4   5   6 . . .

Table 2-1. Relation of n to b and j

The exponential polynomial is calculable for all of the above b except b = 0, in which case the polynomial must be defined. It is all right in that case, because the electron must be measured and defined anyway. But we do not want the electron as part of most higher particles, for we do not want most particle masses to have to be defined. The Universe is designed in such a way that all the particle masses, except the electron, can be calculated from first principles, not needing *ad hoc* definitions. For instance, the η particle has the −++− sub particles. If they were in ground state, the whole particle would have to be measured and defined, not calculated from first principles. But providentially they are all in the 1 state! Thus we can calculate the η mass from first principles. (See later chapters in this book.)

The positional grid has horizontal lines. These separate spaces at different levels in the grid. These represent the different

14

possible orbital energy states a non-dot echon can have. The bottom level is the ground state or 0 state. The next level is the 1 state, and the next level is the 2 state. As higher energy particles are discovered, we will need more energy state levels to explain particle energies. We would simply make more levels in the grid. But three levels explain most common particles.

For example, an electron can have either a minus ½ spin or a plus ½ spin (can be either a - or +) and is in the ground state for a plus or minus semion echon. It is electrically negative. We could write it as either

$$e^-$$
$$\frac{|}{\overline{}}$$
$$\overline{-|}$$

or

$$e^-$$
$$\frac{|}{\overline{}}$$
$$\overline{+|}$$

depending on its spin. Let us just focus on the - spin right now. The same echon could have the higher energy state

$$\mu^-$$
$$\frac{|}{\overline{}}$$
$$\overline{-|}$$
$$|$$

, as a muon,

or the still higher energy state

$$\tau^-$$
$$-|$$
$$\overline{}$$
$$\overline{|}$$
$$|$$

, as a tauon.

This model regards the muon and tauon the same as electrons, except for differences of energy states of the electrinos.

Similarly π^-, K^-, and D^- mesons are related to each other. They are simply composed of negative or positive zero echons in various states.

$$
\pi^+ \qquad \pi^- \qquad K^+ \qquad K^- \qquad D^+ \qquad D^-
$$

$$
\begin{array}{c|c} | \\ \hline | \\ \hline |\,\circ \\ \hline | \end{array} , \quad
\begin{array}{c} | \\ \hline | \\ \hline \circ\,| \\ \hline | \end{array} , \quad
\begin{array}{c} | \\ \hline |\,\circ \\ \hline | \\ \hline | \end{array} , \quad
\begin{array}{c} | \\ \hline \circ\,| \\ \hline | \\ \hline | \end{array} , \quad
\begin{array}{c} |\,\circ \\ \hline | \\ \hline | \\ \hline | \end{array} , \quad
\begin{array}{c} \circ\,| \\ \hline | \\ \hline | \\ \hline | \end{array} .
$$

The bar over a particle symbol indicates it's an antiparticle. But we do not give both the bar and the charge sign. It is like a double negative.

The + or − spin echons can have a minimum of 0 state, and can go up from there. o echons can have a minimum of 1 state, and can go up from there. Recent evidence indicates that zero spin nearly point charge uniton dot yachons can be only in the 2 state, and can never come alone. Often they come with a 0 state + echon. They are always only nearly point charges. Therefore we write them as dots in the 2 level of the grid:

$$
\begin{array}{c}
n \\
\hline |\,\bullet \\ \hline | \\ \hline +\,|
\end{array} , \quad \text{and} \quad
\begin{array}{c}
\bar{n} \\
\hline \bullet\,| \\ \hline | \\ \hline |\,+
\end{array}
$$

Chapter 3

Matter, Antimatter, and Multi-echon Particles

1. Matter, Antimatter

Notice the difference between matter and antimatter—particles and antiparticles: Negatively charged semion echons (minuses or pluses) are matter, whereas positively charged semion echons are antimatter. With quarton echons (o's) it is just the opposite. Positively charged quarton echons (o's) are matter, whereas negatively charged quarton echons are antimatter. Positively charged dots (uniton yachons) are matter, whereas negatively charged dots are antimatter.

2. Multi-echon Particles

We must now begin to figure the composition of multi-echon elementary particles. To begin with, neutrinos seem to be two-component particles. [1] One component is a o and the other a minus or plus:

```
 νe        νe          νμ         ν̄μ        ντ         ν̄τ
  |         |           |          |         ‾|          |‾
  |o  ,   o |    ,    ‾|o   ,    o|‾   ,    _|o   ,    o|_   .
 ‾|        |‾           |          |          |          |
```

_____

[1] Francis Halzen and Alan D. Martin, *QUARKS AND LEPTONS: An Introductory Course in Modern Particle Physics* (New York: John Wiley & Sons, 1984), p. 114.

17

Chapter 4

Spin Hatches, Mass, etc.

1. Spin Hatches

Whereas the positional grid can keep track of the spins of one component particles, it cannot correctly keep track of multi-component particles if they orbit about one another. Some component particles orbit around each other with some spin (whole units of spin—generally spin ± 1). To make our chonomic pictorial equations complete with all the pertinent spin data, we need to add something to the scheme. Let us add an additional hatch below the positional grid to be used for spin numbers:

$$\nu_e$$
$$\frac{\mid}{\mid \circ}$$
$$- \mid$$

$$\underline{1 \mid \tfrac{1}{2}} \quad .$$
$$\mid$$

The top left number in the hatch is the orbital spin (if any) of the component echons. In the case of the electron neutrino above, it is one since the two echons orbit about each other. The top right number in the spin hatch is the total particle spin. It is the sum of the orbital spin (in the top left corner of the hatch) and the total spin shown in the positional grid (intrinsic echon spin). In the above case one echon is an o (has 0 spin) and one echon is a - (has -½ spin). The total intrinsic echon spin is therefore -½. Add that result to the orbital spin (top left hand number in the hatch), and the result is +½—the total expressed in the top right number in the hatch.

Besides the total particle spin, particles in a reaction can contribute or take away angular momentum due to the off-centeredness of their collisions or lines of retreat. This angular momentum is part of the total angular momentum in the system.

18

Total angular momentum is conserved, but particle spin is not always. Therefore we need to keep track of off-centered collision momentum (positional-kinetic angular momentum) in particle equations, and this we will do in the bottom left position of the hatch. The bottom right number in the hatch will be the total spin or angular momentum a particle will contribute to a reaction or carry off from it. It is the total of the particle spin in the top right position of the hatch and the positional-kinetic angular momentum in the bottom left position of the hatch.

2. Mass, Mean Life, Percent Pathway, and Maximum Momentum Per Particle

   The symbolic pictorial scheme lacks one more thing. It lacks a relative indication of the masses of the particles. Let us put a mass value (in MeV) under the particle symbol and above the positional grid in each particle description in our chonomic equations. This will give us a feel for conservation of energy in the reactions. Mean life, percent pathway, and maximum momentum per particle are important particle parameters also. But for simplicity we will not list them in the chonomic pictorial equations. These constants are listed in the headers above the chonomic equations in the appendices in *Electrino Physics* and *Electrino Physics* Draft 2 and in this book Chapter 8.. We will use referenced values in this book. [1] [It is difficult to keep up with the data in this science. The first chapter on this was based on 1990 data; adjustments had to be made with 1992, 1996, 1998, 2002, 2006, 2008 and now 2010 and 2014 data. Physical constants like masses continue to change. However this book is valuable in demonstrating a basic concept.]

---

[1]] J. Beringer *et al* (Particle Data Group) PR **D86**, 010001 (2012) and 2013 partial update for the 2014 edition (URL: http://pdg.lbl.gov).

Chapter 5

Balancing a Simple Reaction

We are almost ready to do particle decay schemes. The plan of balancing the echons and yachons in reactions or decay schemes is that echons and yachons should be conserved in reactions. There should be the same number of positive echons and yachons before and after a reaction. And there should be the same number of negative echons and yachons before and after a reaction. As it turns out, however, the energy state and type of echon or yachon may not be conserved in a reaction. Echons can be bumped up or down in energy states in reactions. Also echons may be fused, changed from − or + to •'s, etc. Yet the total number of negatively charged echons and yachons should be conserved in a reaction, and the total number of positively charged echons and yachons should be conserved in a reaction.

Let us try balancing a simple reaction.

$$\pi^+ \rightarrow e^+ + \nu_e$$

π^+	e^+	ν_e
139.57018	0.510998928	-i0.525 eV

$$
\begin{array}{c}
\pi^+ \\
139.57018 \\
\begin{array}{|c}
\,|\, \\
\,|\,\circ \\
\,|\,
\end{array}
\end{array}
\quad\rightarrow\quad
\begin{array}{c}
e^+ \\
0.510998928 \\
\begin{array}{|c}
\,|\, \\
\,|\, \\
\,|\,-
\end{array}
\end{array}
\quad+\quad
\begin{array}{c}
\nu_e \\
-i0.525\ \text{eV} \\
\begin{array}{|c}
\,|\, \\
\,|\,\circ \\
-\,|\,
\end{array}
\end{array}
$$

$$
\begin{array}{c}
\underline{0\,|\,0} \\
\,|\,
\end{array}
\qquad\qquad
\begin{array}{c}
\underline{0\,|\,-\tfrac{1}{2}} \\
\,|\,
\end{array}
\qquad\qquad
\begin{array}{c}
\underline{1\,|\,\tfrac{1}{2}} \\
\,|\,
\end{array}
$$

The equation as it is written does not balance. Something is missing on the left hand. There are one negatively charged - echon and one positively charged - echon on the right side of the equation that do not appear on the left side of the equation. It is just as though a

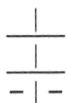

particle is missing on the left hand side of the equation. H. C. Dudley in "Is There an Ether?" [1] theorized there is no such thing as spontaneous radioactive decay, but all radioactive decay results from unseen neutrinos combining with the parent particles according to particle-neutrino cross sections. We also need an unseen particle to combine with the parent particle in our equation. But we don't need a neutrino. A neutrino is a combination of a o echon and a - or + echon. We need the minus-minus particle above. It has spin -1 on the face of the grid. A minus-minus particle may be obtained by having an orbital spin of -1. The total particle spin, then would be -2. Our minus-minus particle that we are missing then would be a graviton with medium mass. Gravitons are plentiful. Apparently our missing particle is a graviton. Except in a quasi particle, however, a

$$
\begin{array}{c}
| \\
\overline{} \\
| \\
\overline{- | -}
\end{array}
$$

is not allowed in this Universe, because such a particle could not have a calculable mass, but would have to be defined, and providence rules that all particles except electrons should be calculable from first principles. The closest particle to that would be

$$
\begin{array}{c}
| \\
\overline{- | -} \\
|
\end{array} \quad ,
$$

which is the lowest state possible graviton. This graviton, however, would not have trivial or negligible mass. All gravitons would have significant mass, which could explain Dark Matter or Dark Energy.

Unseen particles of one kind or another are in decay schemes of all unstable elementary particles. Often more than one particle combines with the parent particle in the decay process. Other particles that we need are

```
      |              • | •              |
    — | —              |              o | o
      |   ,            |   ,            |    .

    1 | 0            1 | 1            1 | 1
      |                |                |
```

The respective particles are a π^0 particle, a photon, and a spin 1 graviton. (There are also spin 2 gravitons.) Pions and photons have been previously known. Spin 1 gravitons are previously unknown. They are plentiful in decay schemes, however.

We have a confusing array of gravitons. Gravitons may be composed of -, +, or o echons, and may be in different energy states. We need compact symbols to differentiate these particles. Let us call gravitons g^{xxyy} where x is the type of echon (-, +, o, or o- or o+), and y is the energy level of the graviton. As an abbreviation we may omit the energy level for lowest state. The following are examples of that terminology:

```
   g⁺            g⁻¹           g°           g°°⁻²
    |             |             |           o | o
  + | +   ,     — | —   ,     o | o   ,       |    .
    |             |             |             |

  1 | 2        -1 | -2        1 | 1        -1 | -1
    |             |             |             |
```

$$g^+ \qquad g^{-1} \qquad g^o \qquad g^{oo-2}$$

Let us now try to go back and balance our decay equation.

$$\pi^+ \rightarrow e^+ + \nu_e \,.$$

```
    π⁺          g⁻          q.p.        e⁺           νₑ
 139.570  70.0          ?          .5109     -i0.525 eV
    |           |           |           |            |
    |           |           |           |            |
   |o  +     -|-    →     |o    →      |      +    |o
    |           |         -|-         |-           -|
```

$$\pi^+ \; (139.570) \quad + \quad g^- \; (70.0) \quad \rightarrow \quad q.p. \; (?) \quad \rightarrow \quad e^+ \; (.5109) \quad + \quad \nu_e \; (-i0.525\,\text{eV})$$

```
  0|0        -1|-2          |          0|-½         1|½
   |           |            |           |            |
               |_____|
             un-observed
             particle
```

We insert an intermediate step in the reaction, a combined particle called q.p. for quasi-particle. The o echon knocks the graviton -- pair to the ground state as it combines with the parent particle to form a jumbo particle which redivides into the daughter particles. This intermediate step will appear more necessary in later decay schemes.

Now let us try to balance the spins in the equation. The echons on the face of the positional grids are balanced, except for energy states. The total particle spins, however, are not balanced. That means there must be some positional-kinetic angular momentum in the particles coming in to or going out of the reaction, or both, to make the total spins balance. We wish to write our decay schemes in the frame of reference of the parent particles. Therefore, we never assign positional-kinetic angular momentum to a parent particle. We can put a 0 in the bottom left hand corner of the spin hatch for the π^+. Adding the spins for the π^+, that means there will be a total 0 spin in the bottom right hand corner of the spin hatch for the π^+ as well.

The graviton probably has +1 positional-kinetic angular momentum, because, in this problem, there is no other incoming particle that can have positional-kinetic angular momentum. And positional-kinetic angular momentum can be at most +1, or at least, -1 per particle. So we can write 1 for the bottom left hatch position for the g^-. The bottom right positions in the hatches are just particle totals, the sum of the upper right numbers and the lower left numbers. Thus there will be a 1 and -1 for the bottom

23

left and bottom right hand positions in the spin hatch for the graviton. Our equation now looks like the following:

$$\pi^+ \rightarrow e^+ + \nu_e \ .$$

π^+	g^-	q.p.	e^+	ν_e
139.5	70.0	?	0.5109	-i0.525 eV

```
   |          |                |            |                    |
 __|o  +    _-_|_-_     →    __|o    →    __|           +      __|o
   |          |               _-_|_-_       |_-                 _-_|

  0|0       -1|-2            |            0|-½               1|½
  0|0        1|-1           _|_           _|                  _|
             |_____|
             un-observed
              particle
```

Now we can find the spin hatch numbers for the quasi particle. Total spin is conserved. So the sum of the bottom right hatch numbers on the left hand side of the equation is the total spin (bottom right hand position of the hatch) of the quasi particle. That number is -1. There is never any positional-kinetic angular momentum for a quasi-particle. So we can immediately fill the bottom left hatch position with a zero. The top right hatch number is the difference between the bottom right and bottom left numbers. So in this instance the top right number is -1. The top right number minus the spin on the face of the positional grid equals the top left number in the spin hatch. In this case we have two - echons and a o echon in the positional grid. The o echon contributes no spin. Each - echon contributes -½ spin, so the total on the face of the positional grid is -1. The top right hand spin hatch number is -1. Therefore the top left hand spin hatch number is 0 net (total particle spin minus intrinsic spin on the face of the positional grid), which we denote as 1-1 (to keep track of oppositely orbiting particles within the system). This is the net orbital spin in the quasi particle. The - echons are orbiting in the - direction with -1 spin. Because of the positional-kinetic angular

24

momentum in the graviton, the minus echons orbit with +1 spin relative to the o echon. Now we can write the quasi particle as

$$
\begin{array}{c}
\text{q.p} \\
| \\
\hline
\quad |\,\text{o} \quad . \\
\hline
-\,|- \\
\\
\underline{1\text{-}1\,|\,\text{-}1} \\
0\,|\,\text{-}1
\end{array}
$$

Now we can try to balance the right hand side of the equation. The total spin of the quasi particle is -1. But the total of the particle spins on the right side of the equation is 0. So we need positional-kinetic angular momentum to balance the equation. The v_e probably should not have positional-kinetic angular momentum because it is a balanced two-component system. Thus we should add a positional-kinetic angular momentum of -1 to the e^+ particle. The equation now balances perfectly.

$$\pi^+ \rightarrow e^+ + v_e \;.$$

π^+	g^-	q.p.	e^+	v_e
139.57	70.0	?	0.5109	-i0.525 eV

$$
\begin{array}{ccccc}
\dfrac{|}{\;|\,\text{o}\;} & + & \dfrac{|}{-\,|-} & \rightarrow & \dfrac{|}{\;|\,\text{o}\;} & \rightarrow & \dfrac{|}{\;|-} & + & \dfrac{|}{\;|\,\text{o}} \\
| & & | & & -\,|- & & & & -\,|
\end{array}
$$

$$
\begin{array}{ccccc}
\dfrac{0\,|\,0}{0\,|\,0} & \dfrac{-1\,|\,\text{-}2}{1\,|\,\text{-}1} & \dfrac{1\text{-}1\,|\,\text{-}1}{0\,|\,\text{-}1} & \dfrac{0\,|\,\text{-}\frac{1}{2}}{-1\,|\,\text{-}1\frac{1}{2}} & \dfrac{1\,|\,\frac{1}{2}}{0\,|\,\frac{1}{2}}
\end{array}
$$

|_____|
un-observed
particle

These pictorial equations are powerful, for they show not only what echons are involved in the reactions, but whether they are

25

upside down or right side up, what energy state they are in, what are their intrinsic spins and orbital spins, and whether or not and in what direction the particles are off-centered in the collisions and recoil paths. The mass numbers above the positional grids show whether particle reactions are possible. In all systems where the second law of thermodynamics is in force, the combined mass of the particles on the left hand side of the equation must be greater than the combined mass of the particles on the right side of the equation, so that energy may be released in kinetic energy and heat to increase entropy. The mass relationships can be seen at a glance with these equations.

Notice in the above equation, echons are conserved. What comes in goes out. That quality does not exist in current quark and lepton models of physics. Echons can construct all leptons. Yachons and echons can construct all hadrons. Each lepton does not have to be a separate elementary particle. In yachons and echons (•, o, +, -), we have found something truly elementary.

---

[1]   H. C. Dudley, *Ind Res.* 43, 44 (Nov. 15, 1974).

Chapter 6

Electrino Process Types

There are hundreds or thousands of possible particle decay schemes. This section will illustrate a few basic types.

A. Wrenching Echons From Orbits to Straight-Forward, Simple Echon Recombinations.

The sample equation in Balancing a Simple Reaction, fits in this category. The following are two more examples.

1. $n \rightarrow p + e^- + \nu_e$

Let us consider magnetism and beta decay. One echon cannot mediate the magnetic force between echons. For a mediating particle, such as the graviton, there must be positive and negative echons with whole orbital spins. When a graviton gravitationally combines with a semion echon, the magnetic moment of the semion echon tries to realign the magnetic moments of the positive and negative orbital echons. At a distance from the semion echon, the gravitons have zero magnetic moments, since the orbital magnetic moments of the positive and negative echons cancel. But as the graviton approaches the semion echon, its magnetic moment tries to align the orbital moments of the positive and negative echons. This is impossible without wrenching the echons out of orbit, which is done by the magnetic force.

Let us review two particles introduced in [1]Chapter 6, the proton and the neutron. Experimental tests seem to indicate that the proton is a three component particle. It is a baryon, and baryon number is conserved in reactions. That is, it is a heavy particle. In all reactions where there is a heavy particle, there will always result a heavy particle after the reaction. The proton must therefore have a different stable constituent than have leptons and mesons which we have studied already (composed of -, +, and o

echons in their various states). A proton, then, must contain a dot
(•) yachon. Try the following for the formula for a proton.

$$
\begin{array}{c}
\text{p} \\
938.272046 \\
\underline{\quad|\ \bullet\ } \\
\underline{\quad\ \ |\ \text{o}\ } \\
+\ | \\
\end{array}
$$

$$
\underline{1\text{--}1\ |\ \tfrac{1}{2}} \quad .
$$
$$
|
$$

The •, o, and + are all in their respective minimum states. The
proton is stable. The + echon orbits about the dot yachon with -1
orbital spin. The o echon orbits about the dot yachon with +1
orbital spin.   Thus we have a total of 1-1 for the orbital spin
number.

The neutron is also a baryon, but has one less positive charge
compared to the proton. It has $\mp$ ½ spin. A proton can trade a π^+
(o echon) with a neutron, converting it into another proton.
Therefore it is logical that the formula of the neutron should be

$$
\begin{array}{c}
\text{n} \\
939.565379 \\
\underline{\quad|\ \bullet\ } \\
\underline{\quad\ \ |\ } \\
+\ | \\
\end{array}
$$

$$
\underline{-1\ |\ -\tfrac{1}{2}} \quad .
$$
$$
|
$$

With these formulae we can balance the decay scheme for an
upside down neutron.

$$n \rightarrow p + e^- + \overline{\nu}_e$$

n	g^o	g-	q.p.	p	e^-	$\overline{\nu}_e$
939.565	70.0	70.0	?	938.2	0.510	+i0.525

eV

```
 _•     _|      _|        _•|      _•|      _|      _|
 _|  +  o|o  +  -|-  →   o|o   →  _|o   +  _|   +  o|
 -|      |       |       --|-      -|      -|       |-
```

1|½	1|1	-1|-2	2-1|-½	1-1|-½	0|-½	1|½
0|½	0|1	0|-2	0|-½	0|-½	0|-½	0|½

```
|_____|
```

un-observed
particles

A g^o graviton and a g⁻ graviton gravitationally combine with an upside down neutron. The dot in the neutron and the - in the neutron go to the upside down proton. The negative - echon in a graviton goes to the electron. The positive - echon and negative o echon from the gravitons go to the electron anti-neutrino with +1 orbital spin. The positive o echon goes to the proton. With two different gravitons, this reaction takes place without any extra positional-kinetic angular momentum. This is the first instance we have seen more than one unobserved particle in a reaction. This will be prevalent in the future.

Straight-forward, simple echon recombinations is the simplest of a variety of electrino processes possible. We will give another example of this category of electrino process.

Before we balance the next decay scheme, let us consider a prevalent graviton—a neutrino orbiting about an anti-neutrino. The graviton can be composed of a - echon orbiting about a o echon in the neutrino, with +1 orbital spin; with a - echon orbiting about a o echon in the anti-neutrino, with +1 orbital spin; and the neutrino orbiting about the anti-neutrino with another +1 orbital spin. The total orbital spin of this graviton is +3, and the total intrinsic spin is -1. Thus the total spin of the particle is +2. The particle is a graviton. We will assign it the symbol g^{o-}. A graviton with opposite intrinsic spins and orbital spins would be g^{o+} with -2 net spin.

$$
\begin{array}{ccc}
g^{o-} & & (\nu_e \qquad \overline{\nu}_e) \\
70.0 & & \text{-i0.525 eV +i0.525 eV} \\
\end{array}
$$

$$
\begin{array}{c}
| \\
\underline{\text{o}\,|\,\text{o}} \\
-\,|\,-
\end{array}
\qquad = \qquad
\begin{array}{c}
| \\
\underline{\quad|\,\text{o}} \\
-\,|
\end{array}
\quad
\begin{array}{c}
| \\
\underline{\text{o}\,|\quad} \\
|\,-
\end{array}
$$

$$
\begin{array}{c}
\underline{3\,|\,2} \\
0\,|\,2
\end{array}
\qquad\qquad\qquad
\begin{array}{c}
\underline{3\,|\,2} \\
0\,|\,2
\end{array}
\quad .
$$

2. $\tau^- \rightarrow \mu^- + \overline{\nu}_\mu + \nu_\tau$

$$
\begin{array}{cccccc}
\tau^- & g^{o+} & q.p. & \mu^- & \overline{\nu}_\mu & \nu_\tau \\
1777.05 & 70.0 & ? & 105.65 & \text{i1.686} & \text{-i14.40}
\end{array}
$$

$$
\begin{array}{c}
+| \\
\underline{\quad} \\
|
\end{array}
+
\begin{array}{c}
| \\
\underline{\text{o}\,|\,\text{o}} \\
+\,|\,+
\end{array}
\rightarrow
\begin{array}{c}
+| \\
\underline{+\text{o}\,|\,\text{o}+} \\
|
\end{array}
\rightarrow
\begin{array}{c}
| \\
\underline{+|\quad} \\
|
\end{array}
+
\begin{array}{c}
| \\
\underline{\text{o}\,|\,+} \\
|
\end{array}
+
\begin{array}{c}
+| \\
\underline{\quad|\,\text{o}} \\
|
\end{array}
$$

$$
\begin{array}{cccccc}
\underline{0\,|\,\tfrac{1}{2}} & \underline{-3\,|\,-2} & \underline{-3\,|\,-1\tfrac{1}{2}} & \underline{0\,|\,\tfrac{1}{2}} & \underline{-1\,|\,-\tfrac{1}{2}} & \underline{-1\,|\,-\tfrac{1}{2}} \\
0\,|\,\tfrac{1}{2} & 0\,|\,-2 & 0\,|\,-1\tfrac{1}{2} & -1\,|\,-\tfrac{1}{2} & 0\,|\,-\tfrac{1}{2} & 0\,|\,-\tfrac{1}{2}
\end{array}
$$

$$
|\underline{\qquad\qquad}|
$$

un-observed
particle

Straight forward recombinations.

B. Echons Knocked To Other Energy States

1. $\mu^- \rightarrow e^- + \bar{\nu}_e + \nu_\mu$

μ^-	g^{o-}	q.p.	e^-	$\bar{\nu}_e$	ν_μ
105.658	70.0	?	.510	$-i0.525eV$	$-i1.642$

```
 |        |         |        |        |        |
-|   +   o|o    →  -o|o   →  |   +   o|   +  -|o
 |       -|-        -|-      -|       |-       |
```

$0\mid-\frac{1}{2}$	$3\mid2$	$3\mid1\frac{1}{2}$	$0\mid-\frac{1}{2}$	$1\mid\frac{1}{2}$	$1\mid\frac{1}{2}$
$0\mid-\frac{1}{2}$	$0\mid2$	$0\mid1\frac{1}{2}$	$1\mid\frac{1}{2}$	$0\mid\frac{1}{2}$	$0\mid\frac{1}{2}$

```
     |_____|
```
un-observed particle

A negative muon gravitationally attracts a g^{o-} graviton, combining with it, making a quasi particle. The echons redivide into an electron, electron anti-neutrino, and a muon neutrino. The orbital angular momentum of the graviton is carried away by the electron anti-neutrino, the muon neutrino, and positional-kinetic angular momentum of the electron. This is a straight forward recombination of echons as we would expect from the quasi particle.

2. $\Lambda \rightarrow p + \pi^-$

Let us introduce a new particle here, the Lambda particle. It is heavier than a neutron and has the same spin as a neutron. In yachon-echon balancing particle decay schemes it seems to have the same yachons and echons as a neutron. It must be a neutron in a more energetic state. It must be a

$$\Lambda$$
$$1115.683$$

$$\frac{\ |\ \bullet\ }{+\ |\ }$$
$$|$$

$$\frac{-1\ |\ -\tfrac{1}{2}}{\ |\ }\ .$$

$$\Lambda \rightarrow p + \pi^-$$

Λ	g^o		q.p.		p	π^-
1115.6	70.0		?		938.27204	139.57018

$$\frac{\ |\ \bullet}{+\ |\ }\ +\ \frac{\ |\ }{o\ |\ o} \rightarrow \frac{\ |\ \bullet}{o\ |\ o} \rightarrow \frac{\ |\ \bullet}{\ |\ o} + \frac{\ |\ }{o\ |\ }$$
$$|\qquad\qquad |\qquad\qquad +\,|\qquad\qquad +\,|\qquad\qquad |$$

$-1|-\tfrac{1}{2}$	$1|1$	$1\text{-}1|\tfrac{1}{2}$	$1\text{-}1|\tfrac{1}{2}$	$0|0$
$0|-\tfrac{1}{2}$	$0|1$	$0|\tfrac{1}{2}$	$0|\tfrac{1}{2}$	$0|0$

$$|\rule{2cm}{0.4pt}|$$
un-observed particle

As a g^o graviton forms a quasi particle with a Λ particle, the 1 state + echon in the Λ is knocked to the 0 state in the quasi-particle. The echons redivide into a p and a π^-. The right side of the equation is too heavy to do this with the neutron as the parent particle, it must start with the Λ particle.

C. Reversal of Orbital Spins

Some decay schemes demonstrate the reflection of orbiting yachons and echons in the collisions of different particles. The following are two examples.

1. $K^+ \rightarrow \pi^+ + \pi^0 + \gamma$

Here we introduce a new particle—a π^0. The π^+ and π^- are composed of o echons. That is all they are. A π^+ is a positively charged o echon, and a π^- is a negatively charged o echon. A π^0 also has about the same mass and also has zero spin, but has 0 charge. To have 0 charge the π^0 must have two oppositely charged echons in it. But the echons cannot be o echons, for the echons must have ± 1 orbital spin or the particles will annihilate. If the echons were o echons, the total particle spin would be ± 1. That is the formula for a g^0 graviton. We need 0 spin for the π^0. The only way we can get that is to make the echon intrinsic spin cancel the echon orbital spin. This can be done if the echons are - or + echons and if the orbital echon spins are +1 or -1, respectively. Besides, the balancing of echon equations indicates that this is the formula for the π^0.

$$\Pi^0$$
$$134.9766$$

```
     |
    ---
   - | -
    ---
     |

    1 | 0
    ---
     |      .
```

Let us now use this information to balance decay scheme 1 above.

```
 K⁺        g⁻¹       Y        q.p.      Π⁺       Π⁰        Y
493.6     70.0       0         ?      139.570  134.976    0
  __        _       • | •     • | •      _        _      • | •
 | o                                                
 ___   +  - | -  +   ___   → - | o-  →   | o  +  - | -  +  ___
  |         _        _         _         _        _        _
  |        | |      | |       | |       | |      | |      | |
```

```
0 | 0    -1 | -2   1 | 1   1-1 | -1    0 | 0    1 | 0    -1 | -1
0 | 0     0 | -2   0 | 1     0 | -1    0 | 0    0 | 0     0 | -1
  |_____|
        un-observed
        particles
```

A graviton or a photon knocks a K^+ echon to the 1 state (π^+) as they combine with it to form a quasi-particle. The echons of the graviton and the yachons of the photon collide, exchanging their orbital spins. Thus a graviton is converted into a π^0 particle. The yachons and echons separate into π^+, π^0, and γ. Again, the right side of the equation is too heavy to start with a π^+, but there is enough mass on the left side of the equation starting with the K^+ particle.

2. $K^+ \rightarrow e^+ + \nu + \pi^0 + \overline{\pi^0}$

K^+	g^-	g^-	g^+	q.p.
493.677	70.0	70.0	70.0	?

$$
\begin{array}{ccccccc}
\dfrac{|o}{|\ \ } & + & \dfrac{|\ }{-|-} & + & \dfrac{|\ }{-|-} & + & \dfrac{|\ }{+|+} & \rightarrow & \dfrac{|\ }{-+|o+-} \rightarrow \\
|\ & & |\ & & |\ & & |\ & & -|-
\end{array}
$$

$\dfrac{0\,|\,0}{0\,|\,0}$	$\dfrac{-1\,|\,-2}{1\,|\,-1}$	$\dfrac{-1\,|\,-2}{0\,|\,-2}$	$\dfrac{1\,|\,2}{0\,|\,2}$	$\dfrac{2-2\,|\,-1}{0\,|\,-1}$

```
      |_____|
      un-observed particles
```

e^+	ν_e	π^0	$\overline{\pi^0}$
0.510998	$-i0.525$eV	134.9766	134.9766

$$
\rightarrow\ \dfrac{|\ }{|\ } + \dfrac{|\ }{|o} + \dfrac{|\ }{-|-} + \dfrac{|\ }{+|+}
$$

$\dfrac{0\,|\,-\frac{1}{2}}{-1\,|\,-1\frac{1}{2}}$	$\dfrac{1\,|\,\frac{1}{2}}{0\,|\,\frac{1}{2}}$	$\dfrac{1\,|\,0}{0\,|\,0}$	$\dfrac{-1\,|\,0}{0\,|\,0}$

A graviton knocks the K^+ echon to the 1 state as two g^- and a g^+ graviton combine with it to form a quasi particle. The echons of a g^+ and a g^- graviton collide in the quasi particle, and reflect, exchanging their orbital spins. They go out as a π^0 and its anti-particle.

D. Flipping Intrinsic Spins

Echons do not always come as --, ++, or oo in positional grids. Every conceivable combination is employed in chonomics, such as -+ or +-, as in the ω(782) employed in the decay schemes below.

1. ω(782) → π⁻ + π⁺ + π⁰

```
ω(782)     g°        q.p.      π⁻      π⁺        π⁰
782.65     70.0        ?      139.57  139.57    134.97
  |         |           |       |       |         |
 -|+ +    o|o  →     -o|o-  →   o|    +  |o  +   -|-
  |         |           |       |       |         |

-1|-1     1|1        1|0      0|0      0|0       1|0
 0|-1     0|1        0|0      0|0      0|0       0|0
   |_____|
un-observed particle
```

A g° graviton combines with an ω(782) to form a quasi-particle. The +1 orbital spin of the graviton flips the + echon of the ω(782) right side up to a - echon. After that, straight-forward recombinations yield π⁻, π⁺, and π⁰. Balancing the spins in this decay scheme is tricky, however. Since the intrinsic spin of an echon is flipped, the intrinsic spins on the face of the grids in this scheme do not balance, as is usually the case. Therefore the orbital spins in the quasi-particle do not balance with the left hand side of the equation either. But the total spins still balance.

```
2.  ω(782) → π⁰ + γ
ω(782)     γ          q.p.              π⁰         γ
782.65     0                          134.97       0
  |      •|•          •|•               |        •|•
 -|+ +    |    →     -|-    →          -|-  +     |
  |       |           |                 |         |

1|1      -1|-1      1-1|-1            1|0       -1|-1
0|1      -1|-2       0|-1             0|0        0|-1
   |_____|
unobserved particle
```

35

A photon with -1 positional-kinetic angular momentum strikes an $\omega(782)$, flipping over its + echon to be a - echon. Straight-forward recombination yields a π^0 and an outgoing photon.

```
3. τ⁻ → ω(782) + π⁻ + ν_τ
    τ⁻          g⁰⁻         q.p.      ω(782)        π⁻        ν_τ
  1776.82     70.0          ?        782.65      139.57   -i14.40
   -|          |           -|          |            |         -|
   _|  +  -o|o-   →   -o|o+   →   -|+   +   o|   +    |o
    |          |            |          |            |          |

   0|-½        3|2         2|1½       1|1          0|0        1|½
   0|-½        0|2         0|1½       0|1          0|0        0|½
    |_____|
  un-observed particle
```

Because of the overall +1 spin of the g^{o-} graviton, the tauon knocks the positive - echon of the g^{o-} graviton upside down to a + echon as the graviton and tauon combine to form a quasi-particle. Straight-forward recombinations result in an $\omega(782)$, π^-, and ν_τ.

E. Fusion of Electrinos

A significant aspect of electrinos is that they may fuse. Elementary particle fusion is not theorized in any other model of physics. In the author's model, electrino fusion has several major impacts.

One is that electrino fusion unites the particles completely. All "elementary" particles can be constructed of quartons, semions, and unitons. But all those are only fusion states of positive and negative octon pairs. Octons have such high charge per mass ratios they can ionize octon pairs out of absolutely nothing. This is significant in the creation and evolution of the Universe. All the Universe could be created from one octon. Quartons could be fused anti-octons, semions fused anti-quartons, and unitons fused anti-semions. The great variety of "elementary" particles can all be

36

constructed from the quartons (in o echons), the semions (in - or + echons) and the unitons (• yachons) and their anti-particles

A second major impact of the fusion of electrinos is that each time the electrinos are fused the particles switch from being matter to antimatter, or vice versa. Current popular models of physics do not predict that, but our model of chonomics shows this already occurs naturally with common particles of physics in laboratories without the physicists' knowledge. When this process is controlled, it is a way of making matter into antimatter (without pair production) for use in annihilating other matter as an energy source and other practical applications.

How do electrinos fuse? There are two semion electrinos in a - or + echon, for example, in an electron. Electrons repel each other because of the Coulomb electric force. But if through high velocity collisions that repulsion is overcome, two electrons with four semions can come so close together that the strong gravitational force overcomes the electric forces, and the four particles all become attracted to each other. When there are only two particles in an orbit, the attractive force keeps the particles in a stable orbit. But where there are four particles all in one orbit, going the same way, that attract each other, they will inevitably not be equally spaced. One semion from one electron and one semion from the other electron will be attracted to each other, and similarly with the other two semions. The four semions will fuse into two unitons, which are whole particles (yachons), which will go their own ways.

Octons fuse to anti-quartons, which fuse to semions, which fuse to anti-unitons. Semion to anti-uniton fusion is fairly easy and occurs naturally. Quarton fusion is more difficult, however, but also occurs naturally. (*See Advanced Electrino Physics*, Draft 3, Chapter 1.) Quartons are normally found in o echons. There are four quartons in a o echon, two orbiting one way and two orbiting the other way. o echons have 0 net spin, therefore they act as bosons, not obeying the Pauli Exclusion Principle, which requires that there be only one particle in each energy state in a system. Boson pions (quarton sources) can be piled one on top of another many times in particle resonances without fusing. It is difficult to

fuse quartons in o echons.   But quartons can be fused rarely
through weak interactions.

The best way to describe natural semion fusion is to show
some chonomic fusion decay schemes, and then to explain them.

1. η'(958) → γ + γ

```
   η'(958)              g⁻                  q.p.1
   957.78              70.0                   ?
     |                   |                     |
  ─────────    +     ───────    →       ─────────────    →
  -o│o-               -│-                  --o│o--
     |                   |                     |

  2-1│0              -1│-2                  2-2│-2
  ─────              ──────                 ───────
   0│0                0│-2                   0│-2
                |_____|
                  unobserved
                  particle
```

```
      q.p.2              g°              γ           γ
        ?                70.0            0           0
     ●●│●●                |            ●│●         ●│●
 →  ───────    →      ───────   +    ──────   +  ──────
     o│o               o│o            │            │
        |                |            |            |

    1-3│-2            -1│-1         -1│-1         1│1
    ──────            ──────        ──────        ────
     0│-2             -1│-2          0│-1          0│1
                |_____|
                 un-observed
                  particle
```

A g⁻ graviton combines with a η'(958) to form a quasi particle.
The four - particles fuse to four dots, forming a second quasi
particle.   The second quasi particle simply redivides into a g°
graviton and two photons.   The photons are observed, but the
graviton is not.

2. $K^0_L \rightarrow \pi^0 + \gamma + \gamma$

The following classic example demonstrates echon recombination, knocking echons to other energy states, collision and reflection of echon spins, and electrino fusion all in one particle decay scheme.

We introduce a new particle here, the K^0-long. There seems to be two different K^0 particles, the K^0-short, and K^0-long—the K^0-short having short decay half lives, and the K^0-long having long half lives. Carefully balancing the echon equations for these two particles indicates that they are composed differently as follows:

```
         K⁰ₛ                    K⁰ₗ
       497.614                497.614
        - | -                  - |
        _____                  _____
         |                      | -
        _____         ,        _____      .
         |                      |


        1 | 0                  1 | 0
        _____                  _____
         |                      |
```

Now let us balance our classic decay scheme:

```
  K⁰ₗ          g⁺          g⁻          q.p.1
497.614       70.0        70.0           ?
 - |           |           |             |
 _____        _____       _____        _____
  | -    +    + | +   +   - | -    →   - - + | + - -   →
 _____        _____       _____        _____
  |            |           |             |


 1 | 0        1 | 2       -1 | -2       2 -1 | 0
 _____        _____       _____       _____
 0 | 0        0 | 2        0 | -2        0 | 0
          |_____|
               un-observed
               particles
```

$$\text{q.p.2} \qquad \bar{\pi}^0 \qquad Y \qquad Y$$
$$? \qquad 134.9766 \qquad 0 \qquad 0$$

First the 2 state – echon in the K^0_L is knocked to the 1 state while two oppositely spinning gravitons combine with the K^0_L to form the first quasi particle. The - echons fuse to dots (q.p.2). One set of the dot echons orbiting with -1 orbital spin collide with the + echons orbiting with +1 orbital spin. The echons reflect and exchange orbital spins. The second quasi particle redivides into an anti- π^0 with + echons and -1 orbital spin and two oppositely spinning photons.

F. Annihilation of Electrinos

Electrinos can also annihilate each other. The electrinos in the following particle systems will annihilate each other, provided they are in the same energy state:

Opposite charged like electrinos with equal and opposite intrinsic spin, no orbital spin, and like energy states annihilate each other. If the energy states are different, they have orbital spin, or their intrinsic spins are not equal and opposite, they will not annihilate so long as they retain their same states. The following are examples of particle systems that will not annihilate so long as the electrinos are in the following states:

```
  o |            |            |            |
  ———          ———          ———          ———
  | o          o | o        + | +        - | +  .
   |    ,       |    ,       |    ,       |

 0 | 0        1 | 1        1 | 0        1 | 1
 ———          ———          ———          ———
   |            |            |            |
```

[1] Gordon L. Ziegler, Electrino Physics Draft 2. Order from Amazon.com or CreateSpace.com.

Chapter 7

Lepton Summary Table and Lepton Decay Schemes

The quark and lepton model of particle physics divides charges in quarks to 2e/3 and e/3. The electrino model of particle physics does not do that. Instead, it divides charges in electrinos to e, e/2, e/4, and e/8. The electrino model of particle physics does not hold that the quark and lepton model of particle physics is correct. Nevertheless, to facilitate cross referencing with the existing data, this volume will employ quark model titles and classifications in the subsequent classification of particles.

The chonomic structures and decay schemes contained in the following material are the author's, but the particle data come from [1]. "In this Summary Table:

"When a quantity has '(S = ...)' to its right, the error on the quantity has been enlarged by the 'scale factor' S, defined as

$$S = \sqrt{\chi^2 / (N - 1)},$$

where N is the number of measurements used in calculating the quantity. We do this when S > 1, which often indicates that the measurements are inconsistent. When S > 1.25, we also show in the Particle Listings an ideogram of the measurements. . . .

"A decay momentum p is given for each decay mode. For a 2-body decay, p is the momentum of each decay product in the rest frame of the decaying particle. For a 3-or-more-body decay, p is the largest momentum any of the products can have in this frame." [2]

To understand how to read and interpret chonomic structures and decay schemes, see Chapters 1-6.

For gravitons, the mass designations in the chonomic representations is 70.0 MeV Experimental values of m^2 for neutrinos is negative, indicating imaginary mass values for neutrinos (See note at end of this Chapter.). The author's model also predicts the mass values of neutrinos will be negative imaginary.

Just as there is more than one way to skin a cat, there is more than one possible way to balance some decay schemes. For instance, the number of required incoming gravitons can vary, in some cases, depending on whether g^o, g^+, and g^- gravitons are employed, without incoming positional-kinetic angular momentum, or g^{o+} and g^{o-} gravitons are employed, with positional-kinetic angular momentum. Also unobserved incoming π^0s could be employed to balance decay schemes, though that is not necessary in this chapter. There are other flexible parameters in these decay schemes. Often the author just had to pick one example to solve the decay scheme. The following decay schemes illustrate a variety of ways to solve the problems. For a more advanced Summary Table, the various decay modes would have to be summed over the various possible pathways, to determine more fundamental quantities, such as populations of various graviton types.

In Chapter 8 are listed state numbers with each chonomic structure. There is not room for them in this chapter. The state levels in this work, if not obvious to the investigator, may be obtained by looking up the particles in Chapter 8.

The following particles are leptons as counted by the Particle Data Group.

LEPTONS

e $\qquad\qquad J = \frac{1}{2}$

Mass m = 0.510998928 ± 0.000000011 MeV
$\quad$ = (548.57990946 ±0.00000022) x 10^{-6} u
$|m_{e+}-m_{e-}|/m < 8$ x 10^{-9}, CL 90%
$|q_{e+}+q_{e-}|/e < 4$ x 10^{-8}

Magnetic moment anomaly
(g-2)/2 = (1159.65218076 ± 0.00000027) x 10^{-6}
$\quad (g_{e+}-g_{e-})/g_{average} = (-0.5±2.1)$x 10^{-12}
Electric dipole moment $d < ($ 10.5 x 10^{-28} e cm, CL = 90%
$\quad$ Mean life $\tau > 4.6$ x 10^{26} years, CL = 90%

43

$$e^-$$
$$0.510998928$$

$$\begin{array}{c} | \\ \underline{|} \\ -\,| \end{array}$$

$$\underline{0\ |\ -\tfrac{1}{2}}$$
$$|$$

$$\mu \qquad\qquad\qquad J = \tfrac{1}{2}$$

Mass m = 105.6583715 ± 0.0000035 MeV
= 0.1134289267 ± 0.0000000029 u
Mean life τ = (2.196911 ± 0.0000022)x 10^{-6} s
$\tau_{\mu+}/\tau_{\mu-}$ = 1.00002 ± 0.00008
$c\tau$ = 658.6384 m
Magnetic moment anomaly (g-2)/2 = (11659209 ± 6) x 10^{-10}
$(g_{\mu+} - g_{\mu-})/g_{average}$ = (-0.11 ± 0.12) x 10^{-8}
Electric dipole moment d = (-0.1 ± 0.9) x 10^{-19} e cm

μ^+ modes are charge conjugates of the modes below.

$$\mu^-$$
$$105.6583715$$

$$\begin{array}{c} \underline{\perp} \\ \underline{\text{-}\perp} \\ | \end{array}$$

$$\underline{0|\text{-}\tfrac{1}{2}}$$
$$|$$

$\underline{\mu^- \text{ DECAY MODE}} \qquad \underline{\text{Fraction } (\Gamma_i/\Gamma) \quad \text{Confidence level} \quad \overset{p}{\text{(MeV/c)}}}$

$$e^- \bar{\nu}_e \nu_\mu \qquad\qquad \approx 100\% \qquad\qquad\qquad 53$$

μ^-	g^{o-}	q.p.	e^-	$\bar{\nu}_e$	ν_μ
105.6...	70.0	?	0.510...	+i0.525eV	−i1.642

```
   µ⁻              g°⁻           q.p.     e⁻          ν̄ₑ          νµ
 105.6...         70.0           ?      0.510...  +i0.525eV   -i1.642

   |               |             |        |           |           |
  -|      +     -o | o-   →   -o | o  →   |     +     |     +    -|
   |               |          -|-       -|         o|-          |o

  0|-½            3|2         3-1|½      0|-½        1|½         1|½
 ──────         ──────       ──────    ──────      ──────      ──────
  0|-½           -1|1         0|½       0|-½        0|½         0|½
     |_____|
     undetected particle
```

The muon knocks the - - echons of the g^{o-} graviton to ground state, then beta decay followed by straight forward recombinations.

$\underline{\mu^- \text{ DECAY MODE}} \qquad \underline{\text{Fraction } (\Gamma_i/\Gamma) \quad \text{Confidence level} \quad \overset{p}{\text{(MeV/c)}}}$

$$e^- \bar{\nu}_e \nu_\mu \gamma \qquad\qquad (1.4 \pm 0.4)\% \qquad\qquad\qquad 53$$

This decay mode is the same as the above, except there is an additional unobserved low energy photon on the left of the reaction, which is energized by the decay process and detected on the right side of the decay process.

μ⁻ DECAY MODE   Fraction (Γ$_i$/Γ)   Confidence level   (MeV/c)

$e^-\bar{v}_e v_\mu e^+ e^-$ $(3.4 \pm 0.4) \times 10^{-5}$ 53

μ^- g^{o-} g^- q.p
105.6...   70.0   70.0

```
  |                |                 |                   |
 -|      +       -o|o-      +       -|-        →        -o|o       →
  |                |                 |                 --|--
```

```
 0|-½          3|2           -1|-2          3-1|-½
 0|-½          0|2            0|-2            0|-½
```

```
          |                                   |
          unobserved particles
```

e^- $\bar{v}_e$ v_μ e^+ e^-
0.510...   -i0.525eV   -i1.642   0.510...   0.510...

```
 |          |         |          |         |
 |    +    o|    +   -|o    +    |    +    |
-|          |-        |         |-        -|
```

```
0|-½       1|½       1|½       0|-½      0|-½
0|-½       0|½       0|½       0|-½      0|-½
```

Two different unobserved gravitons are simultaneously attracted to the muon – econ parent particle, and the muon knocks the - - echons of both gravitons to the ground state, while retaining its state. Beta decay occurs and straight forward recombination of particles.

τ $J = \frac{1}{2}$

Mass m = 1776.82 ± 0.16 MeV

$\left(m_{\tau^+} - m_{\tau^-}\right) / m_{average} < 2.8x10^{-4}$, $CL = 90\%$

Mean life τ = (290.6 ± 1.0) × 10⁻¹⁵ s

cτ = 87.11 μm

Magnetic moment anomaly > -0.052 and < 0.013, CL = 95%

τ^+ modes are charge conjugates of the modes below. "h$^\pm$" stands for $\pi^\pm$ or K$^\pm$, "ℓ" stands for e or μ. "Neutrals" stands for γ's and/or π^0's.

τ DECAY MODE	Fraction (Γ_i/Γ)	Scale factor/ Confidence level	p (MeV/c)
particle$^-$ ≥ 0 neutrals ≥0 K$^0\nu_\tau$ ("1-prong")	(85.35 ± 0.07)%	S = 1.3	--
particle$^-$ ≥ 0 neutrals ≥0K$^0_L\nu_\tau$	(84.71 ± 0.08)%	S = 1.3	--
$\mu^-\overline{\nu}_\mu\nu_\tau$	(17.41 ± 0.04)%	S=1.1	885

```
  τ⁻        g°⁻        q.p.        μ⁻        ν̄_μ     ν_τ <18.2
 1776      70.0                  105.6...  +0.247  -i14.40
  -|         |         -|          |         |        -|
   |    +  -o|o-   →  -o|o-   →   -|    +   o|-   +    |o
   |         |         |          |         |          |

 0|-½   3|2        3-1|½       0|-½      1|½      1|½
 0|-½  -1|1          0|½       0|-½      0|½      0|½
    |_____|
  un-observed particle
```

A graviton composed of a neutrino and an anti-neutrino orbiting about each other gravitationally combine with a tauon, forming a quasi-particle with it. The echons re-divide straight forwardly.

τ DECAY MODE	Fraction (Γ_i/Γ)	Scale factor/ Confidence level	p (MeV/c)
$\mu^-\overline{\nu}_\mu\nu_\tau\gamma$	(3.6 ± 0.4) x 10^{-3}		885

This decay mode is the same as the above, except there is an additional unobserved low energy photon on the left of the reaction, which is energized by the decay process and detected on the right side of the decay process.

τ⁻ DECAY MODE	Fraction (Γ_i/Γ)	Scale factor/ Confidence level	p (MeV/c)
$e^-\overline{\nu}_e\nu_\tau$	[i] (17.83 ± 0.04)%		888

```
 τ⁻           g°⁻      q.p.          e⁻        V̄ₑ<0i      νₜ <18.2
1776.82      70.0                  0.510... -i0.524 eV  -i14.40
 -|            |        -|          |         |          -|
 __      +  -o|o-  →   o|o    →    __    +   o|   +      |o
  |            |        -|-          |         |-          |

 0|-½        3|2      3-1|½        0|-½       1|½        1|½
 0|-½       -1|1       0|½         0|-½       0|½        0|½
  |           |
   un-observed
   particle
```

The - - echons of the graviton are knocked by the tauon to the
ground state in the q.p. particle. Straight-forward recombinations.

τ⁻ DECAY MODE	Fraction (Γ_i/Γ)	Scale factor/ Confidence level	p (MeV/c)
$\tau^- \to \pi^- + \nu_e$	(10.83 ± 0.06)%	S = 1.2	883

```
 τ⁻           g°        q.p.                 π⁻       νₜ <18.2
1776.82      70.0                  139.5...          -i14.40
 -|            |        -|          |                 -|
 __      +   o|o   →   o|o    →    o|    +            |o
  |            |        |           |                   |

 0|-½        1|1      1|½          0|0                1|½
 0|-½        0|1      0|½          0|0                0|½
  |            |
  un-observed particle
```

Straight-forward recombinations.

τ⁻ DECAY MODE	Fraction (Γ_i/Γ)	Scale factor/ Confidence level	p (MeV/c)
$K^-\nu_\tau$	(7.00 ± 0.10) x 10^{-3}	S = 1.1	820

```
  τ⁻          g°           q.p.          K⁻              ν_τ  >0i
1776         70.0                      493.6...           -i14.40
 -|            |           -o|          o|                 -|
 ‾‾|‾     +   ‾o‾|‾o‾   →   ‾‾|‾o‾   →   ‾‾|‾‾     +      ‾‾|‾o‾
   |            |            |            |                 |

0|-½         1|1          1|½          0|0              1|½
0|-½         0|1          0|½          0|0              0|½
              |_____|
             un-observed
               particle
```

The tauon - echon knocks the negative o echon of the $g°$ graviton to the 2 state as the tauon and graviton combine to form a quasi-particle. Thereafter there is straight forward recombinations.

τ DECAY MODE	Fraction (Γ_i/Γ)	Scale factor/ Confidence level	p (MeV/c)
$h^- \geq 1$ neutrals ν_τ	$(37.10 \pm 0.10)\%$	S = 1.2	--
$h^- \pi^0 \nu_\tau$	$(25.95 \pm 0.09)\%$	S = 1.1	878
$\pi^- \pi^0 \nu_\tau$	$(25.52 \pm 0.09)\%$	S = 1.1	878

```
  τ⁻         g°⁻        q.p.        π⁻         π⁰         ν_τ  <18.2
1776..      70.0                  139.5... 134.9...       -i14.40
 -|           |          -|         |          |           -|
 ‾‾|‾    +  ‾-o‾|‾o-‾ → ‾-o‾|‾o-‾ → ‾o‾|‾  +  ‾-‾|‾-‾  +  ‾‾|‾o‾
   |           |          |          |          |           |

0|-½        3|2        3-1|½       0|0        1|0         1|½
0|-½       -1|1         0|½        0|0        0|0         0|½
              |_____|
      un-observed particle
```

A π^0 is a system of an anti-muon and an muon in orbit about one another with the orbital spin oppositely directed to the intrinsic spin of the anti-muon and muon. A π^0 is not composed of the same echons as $\pi^{\pm}$ particles, though they have similar mass. Decay schemes of π^0 and $\pi^{\pm}$ will bear this out.

τ DECAY MODE	Fraction (Γ_i/Γ)	Scale factor/ Confidence level	p (MeV/c)
$\pi^-\pi^0\text{non-}\rho(770)\nu_\tau$	$(3.0 \pm 3.2)\text{x }10^{-3}$		878
$K^-\pi^0\nu_\tau$	$(4.29 \pm 0.15)\text{x }10^{-3}$		814

τ^-	g^{o-}	q.p.	K^-	π^0	ν_τ
1776..	70.0		493.6...	134.9...	-i14.40

```
 -|         |       -o|        o|         |         -|
 _|  + -o|o- → -|o-  →  _|  + -|-   +   |o
  |         |         |         |         |          |
```

$0\mid-\tfrac{1}{2}$	$3\mid2$	$3-1\mid\tfrac{1}{2}$	$0\mid0$	$1\mid0$	$1\mid\tfrac{1}{2}$
$0\mid-\tfrac{1}{2}$	$-1\mid1$	$0\mid\tfrac{1}{2}$	$0\mid0$	$0\mid0$	$0\mid\tfrac{1}{2}$

un-observed particle

A g^{o-} graviton gravitationally combines with a tau particle. The negative o echon gets knocked up to the 2 state in the process. Straight forward recombinations result in a K^-, π^0, and a ν_τ.

τ DECAY MODE	Fraction (Γ_i/Γ)	Scale factor/ Confidence level	p (MeV/c)
$h^- \geq 2\pi^0 \nu_\tau$	$(10.87 \pm 0.11)\%$	$S = 1.2$	--
$h^-2\pi^0\nu_\tau$	$(9.52 \pm 0.11)\%$	$S = 1.1$	862
$h^-2\pi^0\nu_\tau(\text{ex.}K^0)$	$(9.36 \pm 0.11)\%$	$S = 1.2$	862
$\pi^-2\pi^0\nu_\tau(\text{ex.}K^0)$	$(9.30 \pm 0.11)\%$	$S = 1.2$	862

τ^-	g^{o-}	g^-	q.p.	π^-	π^0	π^0	$\nu_\tau<18.2$
1776..	70.0	70.0		139...	134...	134...	-i14.40

```
 -|       |       |      -|      |       |       |       -|
 _|  + -o|o- +  -|-  → --o|o-- → o|  + -|- + -|- +   |o
  |       |       |       |      |       |       |        |
```

$0\mid-\tfrac{1}{2}$	$3\mid2$	$-1\mid-2$	$4-1\mid\tfrac{1}{2}$	$0\mid0$	$1\mid0$	$1\mid0$	$1\mid\tfrac{1}{2}$
$0\mid-\tfrac{1}{2}$	$1\mid3$	$0\mid-2$	$0\mid\tfrac{1}{2}$	$0\mid0$	$0\mid0$	$0\mid0$	$0\mid\tfrac{1}{2}$

un-observed particles

The system experiences straight-forward recombinations.

τ DECAY MODE	Fraction (Γ_i/Γ)	Scale factor/ Confidence level	p (MeV/c)
$K^-2\pi^0\nu_\tau(\text{ex.}K^0)$	$(6.5 \pm 2.3)\text{ x }10^{-4}$		796

```
  τ⁻        g⁻        g⁰⁻              q.p.         K⁻        π⁰        π⁰    ν_τ  <18.2
1776..     >0i       >0i                          493...    134...    134...  -i14.40
 -|         |         |                -o|         o|        |         |      -|
 _|_   +   _|_   +  _o|o_    →      __|o__   →    _|_   +   _|_   +   _|_   +  _|o
  |         |         |                 |          |        |         |       |

0|-½      -1|-2      3|2              4-1|½        0|0       1|0       1|0      1|½
0|-½       1|-1      0|2               0|½         0|0       0|0       0|0      0|½
           |_____|
              un-observed particles
```

One o echon is knocked to the +2 state by the tauon – echon, as a g⁰⁻ graviton and g⁻ graviton gravitationally combine with a τ⁻ particle. Thereafter it is straight-forward recombinations.

		Scale factor/	p
τ DECAY MODE	Fraction (Γ_i/Γ)	Confidence level	(MeV/c)
$h^- \geq 3\pi^0 \nu_\tau$	$(1.35 \pm 0.07)\%$	$S = 1.$	--
$h^- 3\pi^0 \nu_\tau$	$(1.19 \pm 0.07)\%$		836
$\pi^- 3\pi^0 \nu_\tau (\text{ex.} K^0)$	$(1.05 \pm 0.07)\%$		836

```
  T⁻        g⁰⁻          g⁻          g⁻              q.p.
1776..     70.0         70.0        70.0
 -|         |            |           |              -|
 _|_   +  _o|o_   +    _|_   +     _|_    →     ___o|o___   →
  |         |            |           |              |

0|-½       3|2         -1|-2       -1|-2          3-2|-2½
0|-½       0|2          0|-2        0|-2           0|-2½
           |_____|
                   un-observed particles
```

```
   π⁻          π⁰           π⁰           π⁰          ν_τ  <18.2
139...       134...       134...       134...        -i14.40
  |            |            |            |            -|
 o|    +     _|_    +     _|_    +     _|_    +      _|o
  |            |            |            |            |

0|0          1|0          1|0          1|0          1|½
0|0         -1|-1        -1|-1        -1|-1          0|½
```

51

A g^{0-} graviton and two g^- gravitons gravitationally combine with the tauon, forming a quasi-particle. Considering the spins, it is straight-forward recombinations.

				Scale factor/	p
τ^- **DECAY MODE**		Fraction (Γ_i/Γ)	Confidence level		(MeV/c)

$K^-3\pi^0\nu_\tau(\text{ex.}K^0)$ $(4.8 \pm 2.2) \times 10^{-4}$ 765

```
 τ⁻        g⁰⁻        g⁻         g⁻          q.p.
1776..   70.0       70.0       70.0
 -|        |          |          |          -o|
 _|_  +  _-o|o-_  +  _-|-_  +  _-|-_   →  _---|o---_   →
  |        |          |          |           |


0|-½     3|2       -1|-2      -1|-2      3-2|-2½
0|-½     0|2        0|-2       0|-2       0|-2½
          |                                        |
          |_____|
              un-observed particles
 K⁻        π⁰         π⁰         π⁰       νₜ  <18.2
493...    134...     134...     134...      -i14.40
 o|        |          |          |          -|
 _|_  +  _-|-_  +  _-|-_  +  _-|-_   +    _|o_
  |        |          |          |          |


0|0      1|0        1|0        1|0        1|½
0|0     -1|-1      -1|-1      -1|-1       0|½
```

A g^{0-} graviton and two g^- gravitons gravitationally combine with the tauon, forming a quasi-particle. The tauon knocks one o echon to the +2 state. Considering the spins, it is straight-forward recombinations.

			Scale factor/	p
τ^- **DECAY MODE**	Fraction (Γ_i/Γ)		Confidence level	(MeV/c)
$h^-4\pi^0\nu_\tau(\text{ex.}K^0)$	$(1.6 \pm 0.4) \times 10^{-3}$			800
$h^-4\pi^0\nu_\tau(\text{ex.}K^0,\eta)$	$(1.1 \pm 0.4) \times 10^{-3}$			800
$K^- \geq 0\pi^0 \geq 0K^0 \nu_\tau$	$(1.572\pm 0.033)\%$		S=1.1	820
$K^- \geq 1 (\pi^0 \text{ or } K^0) \nu_\tau$	$(8.72 \pm 0.32) \times 10^{-3}$		S=1.1	--

Modes with K^0's

K^0_S(particles)$^-\nu_\tau$ (9.2± 0.4) x 10^{-3} S = 1.5   --

$h^-\overline{K}^0\nu_\tau$ (1.00 ± 0.05)%   S = 1.8   812

$\pi^-\overline{K}^0$ (5.4 ± 2.1)x 10^{-4} 812
(non-K*(892)$^-$)ν_τ

```
τ⁻            g⁰⁺          q.p.         π⁻           K⁰ˢ         ντ <18.2
1776..        70.0                      139.5...     497.6...    -i14.40
-|             |           -+|+          |           +|+          -|
 |    +    +o|o+    →      o|o    →     o|     +      |     +       |o
 |             |           o|o          |            |            |

0|-½          -3|-2        -3|-2½  0|0   -1|0         1|½
0|-½          0|-2         0|-2½  -1|-1  -1|-1        -1|-½
       |_____|
     un-observed particle
```

A g^{o+} graviton (a neutrino-anti-neutrino graviton) is gravitationally attracted to a tauon, forming a quasi-particle, knocking the + + pair to the 2 state. Thereafter it is straight-forward recombinations. Anti-K^0_S is when both negative and positive + echons are in the +2 state. Anti-K^0_L is when one + echon is in the +2 state, and one is in the 1 state.

	Scale factor/	p
τ DECAY MODE Fraction (Γ_i/Γ)	Confidence level	(MeV/c)

$K^-K^0\nu_\tau$ (1.59± 0.16)x 10^{-3} 737

```
τ⁻          g⁻      g⁰      q.p.        K⁻          K⁰ᴸ        ντ <18.2
1776..      70.0    70.0                493.6...    497.6...   -i14.40
-|           |       |      --o|         o|          -|          -|
 |    +     -|-  +  o|o  →    |o-  →      |    +      |-    +      |o
 |           |       |        |          |           |           |

0|-½        -1|-2   1|1    1-1|-1½  0|0   1|0         1|½
0|-½        0|-2    0|1    0|-1½   -1|-1  -1|-1        0|½
     |_____|
       un-observed particles
```

53

A g⁻ graviton and a g° graviton are gravitationally attracted to a tauon, forming a quasi-particle with it. The tauon knocks one o echon and one - echon of the gravitons to the +2 state. Thereafter it is straight forward recombinations.

	Scale factor/	p
τ DECAY MODE	Fraction (Γ_i/Γ) Confidence level	(MeV/c)

$h^-\overline{K}^0\pi^0\nu_\tau$ $(\,5.6 \pm 0.4\,)$x 10^{-3} 794

$\pi^-\overline{K}^0\pi^0\nu_\tau$ $(\,4.0 \pm 0.4\,)$x 10^{-3} 794

```
τ⁻          g°⁻     g⁻      q.p.        π⁻       K⁰      π⁰      ν̄τ<18.2
1776..      70.0    70.0                139...   497...  134...  -i14.40
 -|           |       |       --|-        |        -|-      |       -|
 _|   +  -o|o- +  -|-   →   -o|o-   →    o|   +   _|  +  -|-  +   |o
  |           |       |        |          |         |       |        |

0|-½        3|2    -1|-2    3-1|-½      0|0      1|0     1|0     1|½
0|-½        0|2     0|-2    0|-½        0|0      0|0    -1|-1    0|½
  |                   |
     un-observed particles
```

A g⁰⁻ graviton and a g⁻ graviton combine with a tauon, forming a quasi-particle. The tauon knocks both of one graviton - echons to the +2 state. Thereafter it is straight forward recombinations.

	Scale factor/	p
τ DECAY MODE	Fraction (Γ_i/Γ) Confidence level	(MeV/c)

$\overline{K}^0\rho^-\nu_\tau$ $(\,2.2 \pm 0.5\,)$x 10^{-3} 612

```
τ⁻       g⁺     g⁻      g°      q.p.       K̄⁰     ρ(770)⁻  ντ <18.2
1776     70.0   70.0    70.0              497...  775.4    -i14.40
 -|        |      |       |       -+|+      +|+      |        -|
 _|  +  +|+ +  -|-  +   o|o   →  -o|o-  →   _|  +  -o|-  +    |o
  |        |      |       |        |         |        |         |

0|-½     1|2   -1|-2    1|1     2-1|½      -1|0    1-1|-1    1|½
0|-½     0|2    0|-2    0|1     0|½         1|1    0|-1      0|½
  |                      |
     un-observed particles
```

54

A g^+, a g^-, and a g^o graviton combine with a tauon to form a quasi-particle. The tauon knocks both + echons to the 2 state. Otherwise it is straight-forward recombinations.

				Scale factor/	p
τ^- **DECAY MODE**			Fraction (Γ_i/Γ)	Confidence level	(MeV/c)

$$K^-K^0\pi^0\nu_\tau \qquad (1.59\pm 0.20)\text{x }10^{-3} \qquad\qquad 685$$

```
  τ⁻        g-      g⁰⁻     q.p.        K⁻      K⁰      π⁰      ν_τ<0i
1776..     >0i     >0i             493... 497... 134...-i14.40
-|          |       |     --o|-      o|      -|-     |       -|
___         |       |     ____      __      ___     |       ___
 |    +  -|- +  -o|o-  →  -|o-  →   |  +  |  + -|- + |o
 |          |       |       |        |       |       |       |

0|-½     -1|-2    3|2    3-1|-½    0|0     1|0     1|0     1|½
0|-½      0|-2    0|2     0|-½     0|0     0|0    -1|-1    0|½
 |_____|
     un-observed particles
```

A g^- and a g^{o-} graviton combine with a tauon to form a quasi-particle. The tauon knocks a negative o and both - echons from one graviton to the +2 state. Otherwise it is straight-forward recombinations.

				Scale factor/	p
τ^- **DECAY MODE**			Fraction (Γ_i/Γ)	Confidence level	(MeV/c)

$$\pi^-\overline{K}^0\pi^0\pi^0\nu_\tau \qquad (2.6\ \pm 2.4\ )\text{x }10^{-4} \qquad\qquad 763$$

```
  τ⁻           g⁰⁻        g⁻        g⁺        q.p.
1776...       70.0       70.0      70.0
-|             |          |         |        -+|+
___            |          |         |        _____
 |     +   -o|o-   +   -|-   +  +|+  →   --o|o--   →
 |             |          |         |         |

0|-½         3|2       -1|-2      1|2     4-1|1½
0|-½         0|2        0|-2      0|2      0|1½
 |_____|
         unobserved particles
```

55

π⁻ $\quad$ $\overline{K}^0$ $\quad$ π⁰ $\quad$ π⁰ $\quad$ v_τ <18.2

$$
\begin{array}{ccccc}
\pi^- & \overline{K}^0 & \pi^0 & \pi^0 & v_\tau \, <18.2 \\
139\ldots & 497\ldots & 134\ldots & 134\ldots & -i14.40
\end{array}
$$

$$
\frac{\begin{array}{c}| \\ \circ\,| \\ |\end{array}}{} \;+\; \frac{\begin{array}{c}+\,|\,+ \\ | \\ |\end{array}}{} \;+\; \frac{\begin{array}{c}| \\ -\,|\,- \\ |\end{array}}{} \;+\; \frac{\begin{array}{c}| \\ -\,|\,- \\ |\end{array}}{} \;+\; \frac{\begin{array}{c}-\,| \\ | \\ |\,\circ\end{array}}{}
$$

$$
\frac{0\,|\,0}{0\,|\,0} \qquad \frac{-1\,|\,0}{0\,|\,0} \qquad \frac{1\,|\,0}{0\,|\,0} \qquad \frac{1\,|\,0}{1\,|\,1} \qquad \frac{1\,|\,\tfrac{1}{2}}{0\,|\,\tfrac{1}{2}}
$$

The decay mode employs three gravitons combining with the tauon to form a quasi-particle. There are massive particle collisions and reflecting the orbital spins in the quasi-particle. The + + echons of the g^+ graviton are knocked to the 2 state. Otherwise it is straight-forward recombinations.

		Scale factor/	p
τ DECAY MODE	Fraction (Γ_i/Γ)	Confidence level	(MeV/c)
$\pi^- K^0 \overline{K}^0 v_\tau$	(1.7± 0.4) x 10^{-3}	S = 1.8	682

$$
\begin{array}{cccccccc}
\tau^- & g^+ & g^{o-} & q.p. & \pi^- & K^0 & \overline{K}^0 & v_\tau<18.2 \\
1776..& 70.0 & 70.0 & & 139\ldots & 497\ldots & 497\ldots & -i14.40
\end{array}
$$

$$
\frac{\begin{array}{c}-| \\ | \\ |\end{array}}{} + \frac{\begin{array}{c}| \\ +|+ \\ |\end{array}}{} + \frac{\begin{array}{c}| \\ -\,\circ|\circ- \\ |\end{array}}{} \rightarrow \frac{\begin{array}{c}--+|+- \\ \circ|\circ \\ |\end{array}}{} \rightarrow \frac{\begin{array}{c}| \\ \circ| \\ |\end{array}}{} + \frac{\begin{array}{c}-|- \\ | \\ |\end{array}}{} + \frac{\begin{array}{c}+|+ \\ | \\ |\end{array}}{} + \frac{\begin{array}{c}-| \\ | \\ |\circ\end{array}}{}
$$

$$
\frac{0\,|-\tfrac{1}{2}}{0\,|-\tfrac{1}{2}} \;\; \frac{1\,|\,2}{-1\,|\,1} \quad \frac{3\,|\,2}{0\,|\,2} \qquad \frac{4-1\,|\,2\tfrac{1}{2}}{0\,|\,2\tfrac{1}{2}} \quad \frac{0\,|\,0}{1\,|\,1} \quad \frac{1\,|\,0}{1\,|\,1} \quad \frac{-1\,|\,0}{0\,|\,0} \quad \frac{1\,|\,\tfrac{1}{2}}{0\,|\,\tfrac{1}{2}}
$$

```
        |_____|
        un-observed particles
```

A g^+ and a g^{o-} (an upside down g^{o+}) graviton combine with a tauon to form a quasi-particle. The + and - echons from the gravitons are knocked to the +2 state. The + echons trade orbital spin and positional-kinetic angular momentum to form an anti-K^0 particle. Thereafter it is straight-forward recombinations.

τ DECAY MODE	Fraction (Γ_i/Γ)	Scale factor/ Confidence level	p (MeV/c)
$\pi^- K^0_S K^0_S \nu_\tau$	$(2.31 \pm 017) \times 10^{-4}$	S= 1.9	682

τ^-	g^{o-}	g^-	q.p.
1776..	70.0	70.0	

$$\begin{array}{c} - \mid \\ \underline{\mid} \\ \mid \end{array} + \begin{array}{c} \mid \\ \underline{-o\mid o-} \\ \mid \end{array} + \begin{array}{c} \mid \\ \underline{-\mid-} \\ \mid \end{array} \rightarrow \begin{array}{c} ---\mid-- \\ \underline{o\mid o} \\ \mid \end{array} \rightarrow$$

$0 \mid -\frac{1}{2}$	$3 \mid 2$	$-1 \mid -2$	$3-1 \mid -\frac{1}{2}$
$0 \mid -\frac{1}{2}$	$0 \mid 2$	$0 \mid -2$	$0 \mid -\frac{1}{2}$

un-observed particles

π^-	K^0_S	K^0_S	ν_τ <18.2
139...	497...	497...	-i14.40

$$\begin{array}{c} \mid \\ \underline{o\mid} \\ \mid \end{array} + \begin{array}{c} -\mid- \\ \underline{\mid} \\ \mid \end{array} + \begin{array}{c} -\mid- \\ \underline{\mid} \\ \mid \end{array} + \begin{array}{c} -\mid \\ \underline{\mid o} \\ \mid \end{array}$$

$0 \mid 0$	$1 \mid 0$	$1 \mid 0$	$1 \mid \frac{1}{2}$
$-1 \mid -1$	$0 \mid 0$	$0 \mid 0$	$0 \mid \frac{1}{2}$

The tauon knocks the graviton - echons to the +2 state. Subsequently there are straight forward recombinations.

τ DECAY MODE (MeV/c)	Fraction (Γ_i/Γ)	Scale factor/ Confidence level	p
$\pi^- K^0_S K^0_L \nu_\tau$	$(1.2 \pm 0.4) \times 10^{-3}$	S = 1.78	682

```
 τ⁻          g°⁻          g⁻              q.p.
1776..      70.0        70.0
 -|           |            |             ---|-
 _|_  +    -o|o-  +     -|-      →      _o|o-_    →
  |           |            |              |

 0|-½        3|2         -1|-2           3-1|-½
 0|-½        0|2         0|-2            0|-½
    |_____|
         un-observed particles
```

```
  π⁻          K⁰_S         K⁰_L        ν_τ <18.2
139...       497...       497...       -i14.40
  |           -|-          -|            -|
 o|_  +       _|_  +      _|-_  +       _|o_
  |            |            |             |

 0|0         1|0          1|0           1|½
 -1|-1       0|0          0|0           0|½
```

The tauon knocks all but one of the graviton - echons to the +2 state. Then there are straight forward recombinations.

	Fraction (Γ_i/Γ)	Scale factor/ Confidence level	p (MeV/c)
τ DECAY MODE			
$\pi^- K^0_S K^0_L \pi^0 \nu_\tau$	$(3.1 \pm 1.2) \times 10^{-4}$		614

```
 τ⁻          g°⁻         g⁻          g⁻              q.p.
1776..      70.0        70.0        70.0
 -|           |           |            |             ---|-
 _|_  +    -o|o-  +     -|-  +      -|-      →      -o|o--    →
  |           |           |            |              |

 0|-½        3|2        -1|-2        -1|-2           3-2|-2½
 0|-½        0|2        0|-2         0|-2            0|-2½
    |_____|
         un-observed particles
```

58

π⁻   K⁰_S   K⁰_L   π⁰   ν_τ <18.2

π^- K^0_S K^0_L π^0 $\nu_\tau <18.2$

139...   497...   497...   134...   -i14.40

0 | 0	1 | 0	1 | 0	1 | 0	1 | ½
0 | 0	-1 | -1	-1 | -1	-1 | -1	0 | ½

Three un-observed gravitons combine with a tauon to form a quasi-particle. The tauon knocks three of the - echons to the 2 state.. The other graviton - echons remain in 1 state. Subsequently there are straight forward recombinations.

τ DECAY MODE	Fraction (Γ_i/Γ)	Scale factor/ Confidence level	p (MeV/c)
$K^0 h^+ h^- h^- \nu_\tau$	$(2.3 \pm 2.0) \times 10^{-4}$		760

Modes with three charged particles

$h^- h^- h^+ \geq 0$neut. ν_τ("3-prong")	$(15.20 \pm 0.08)\%$	S = 1.3	861
$h^- h^- h^+ \geq 0$neutrals ν_τ (ex. $K^0_S \to \pi^+\pi^-$)	$(14.57 \pm 0.07)\%$	S = 1.3	861
$h^- h^- h^+ \nu_\tau$	$(9.80 \pm 0.07)\%$	S = 1.2	861
$h^- h^- h^+ \nu_\tau$(ex.K^0)	$(9.46 \pm 0.06)\%$	S = 1.2	861
$h^- h^- h^+ \nu_\tau$(ex.K^0,ω)	$(9.42 \pm 0.06)\%$	S = 1.2	861
$\pi^- \pi^- \pi^+ \nu_\tau$	$(9.31 \pm 0.06)\%$	S = 1.1	861

τ^- g^0 g^0 q.p. π^- π^+ π^- $\nu_\tau <18.2$

1776..   70.0   70.0   139...   139...   139...   -i14.40

0 | -½	1 | 1	-1 | -1	1 -1 | -½	0 | 0	0 | 0	0 | 0	1 | ½
0 | -½	0 | 1	0 | -1	0 | -½	0 | 0	-1 | -1	0 | 0	0 | ½

un-observed particles

Straight-forward recombinations.

τ DECAY MODE	Fraction (Γ_i/Γ)	Scale factor/ Confidence level	p (MeV/c)
$\pi^-\pi^+\pi^-\nu_\tau(\text{ex.}K^0)$	(9.02± 0.06)%	S = 1.1	861
$\pi^-\pi^+\pi^-\nu_\tau(\text{ex.}K^0,\omega)$	(8.99± 0.06)%	S = 1.1	861
$h^-h^-h^+\geq 1\text{neutrals }\nu_\tau$	(5.39± 0.07)%	S = 1.2	--
$h^-h^-h^+\geq 1\pi^0\nu_\tau(\text{ex. }K^0)$	(5.09± 0.06)%	S = 1.2	--
$h^-h^-h^+\pi^0\nu_\tau$	(4.76± 0.06)%	S = 1.2	834
$h^-h^-h^+\pi^0\nu_\tau(\text{ex.}K^0)$	(4.59± 0.06)%	S = 1.2	834
$h^-h^-h^+\pi^0\nu_\tau(\text{ex.}K^0,\omega)$	(2.79± 0.08)%	S = 1.2	834
$\pi^-\pi^+\pi^-\pi^0\nu_\tau$	(4.62± 0.06)%	S = 1.2	834

```
τ⁻         g°⁻         g°                    q.p.
1776..     70.0        70.0
 -|          |           |              -|
 _|    +   -o|o-   +   o|o      →     -oo|oo-    →
  |          |           |               |


0|-½        3|2        -1|-1           3-1|½
0|-½        0|2         0|-1            0|½
          |                      |
       un-observed particles


 π⁻        π⁺         π⁻           π⁰       ν_τ  <18.2
139...    139...     139...       134...   -i14.40
  |          |          |            |        -|
 o|    +    |o    +   o|     +     -|-   +    |o
  |          |          |            |         |


 0|0        0|0        0|0          1|0       1|½
 0|0        0|0        0|0          0|0       0|½
```

A g^{o-} graviton and a g^o graviton are gravitationally attracted to a tauon. They form a quasi-particle with it. Then there are straight-forward recombinations.

τ^- **DECAY MODE**	Fraction (Γ_i/Γ)	Scale factor/ Confidence level	p (MeV/c)
$\pi^-\pi^+\pi^-\pi^0\nu_\tau(\text{ex.K}^0)$	(4.48± 0.06)%	S = 1.2	834
$\pi^-\pi^+\pi^-\pi^0\nu_\tau(\text{ex.K}^0,\omega)$	(2.70± 0.08)%	S = 1.2	834
$h^-h^-h^+2\pi^0\nu_\tau$	(5.08 ± 0.32)x 10^{-3}		797
$h^-h^-h^+2\pi^0\nu_\tau(\text{ex.K}^0)$	(4.98 ± 0.32)x 10^{-3}		797
$h^-h^-h^+2\pi^0\nu_\tau(\text{ex.K}^0,\varpi,\eta)$	(1.0 ±0. 4)x 10^{-3}		797

```
  τ⁻        g°        g°       g⁻      g⁻       q.p.
1776..    70.0      70.0     70.0    70.0

 -|         |         |        |       |         -|
 _|   +   o|o   +   o|o   +  -|-  +  -|-  →   --oo|oo--  →
  |         |         |        |       |          |

0|-½      1|1       1|1      -1|-2  -1|-2     2-2|-2½
0|-½      0|1       0|1      0|-2   0|-2       0|-2½
           |    un-observed particles       |

  π⁻       π⁻        π⁺       π⁰      π⁰      ν_τ  <18.2
139...   139...    139...  134...  134...     -i14.40

 |         |         |        |       |          -|
o|   +   o|   +    |o   +  -|-  +  -|-  +      |o
 |         |         |        |       |          |

0|0       0|0       0|0      1|0     1|0       1|½
0|0      -1|-1      0|0     -1|-1   -1|-1      0|½
```

Two g^0 gravitons and two g^- gravitons gravitationally combine with the tauon, forming a quasi-particle with it. Subsequently there are straight-forward recombinations.

$h^-h^-h^+3\pi^0\nu_\tau$ (2.3 ± 0.6)x 10^{-4} S = 1.2   749

```
  τ⁻          g°⁻         g°⁻         g⁻              q.p.
1776..       70.0        70.0        70.0
 -|            |           |           |                 -|
 _|_   +    -o|o-   +   -o|o-   +    -|-     →    ---oo|oo---   →
  |            |           |           |                  |

0|-½         3|2         3|2         -1|-2           6-1|1½
0|-½         0|2         0|2          0|-2             0|1½
         |un-observed particles |
```

```
   π⁻         π⁻          π⁺          π⁰          π⁰          π⁰    ν_τ <18.2
 139..      139..       139..       134..       134..       134..   -i14.40
   |           |           |           |           |           |       -|
 _o|_   +   _o|_   +    _|o_   +    -|-   +    -|-   +    -|-     _|o_
   |           |           |           |           |           |       |

 0|0        0|0         0|0         1|0         1|0         1|0       1|½
 0|0        0|0         0|0         1|1         0|0         0|0       0|½
```

Outgoing positional-kinetic angular momentum makes this decay mode possible. Three gravitons join the tauon in forming a quasi-particle. With graviton echon collisions and reflecting of orbital spins, the particles have straight forward recombinations.

		Scale factor/	p
τ **DECAY MODE**	Fraction (Γ_i/Γ)	Confidence level	(MeV/c)
$K^-h^+h^- \geq 0$neutrals ν_τ	(6.35 ± 0.24)x 10^{-3}	S = 1.5	794
$K^-\pi^+\pi^- \geq 0$neutrals ν_τ	(4.38 ± 0.19)x 10^{-3}	S = 2.7	794
$K^-\pi^+\pi^-\nu_\tau$	(3.49 ± 0.16)x 10^{-3}	S = 1.9	794
$K^-\pi^+\pi^-\nu_\tau$(ex.K^0)	(2.94 ± 0.15)x 10^{-3}	S = 2.2	794
$K^-\pi^+\pi^-\pi^0\nu_\tau$	(1.35 ± 0.14)x 10^{-3}		763
$K^-\pi^+\pi^-\pi^0\nu_\tau$(ex.$K^0$)	(8.1 ± 1.2)x 10^{-4}		763

```
   τ⁻            g°⁻           g°            q.p.
1776...        70.0          70.0                         
 -|             |             |             -o|
 ___    +     _____  +    _____    →    _____   →
  |          -o|o-          o|o            -o|oo-
  |             |             |               |

 0|-½          3|2          -1|-1          3-1|½
 ____          ___          _____          _____
 0|-½          0|2          0|-1           0|½
                |_____|
                   un-observed particles
```

```
  K⁻          π⁺           π⁻           π⁰          ν_τ <18.2
 493...      139...       139...       134...       -i14.40
 o|           |            |            |            -|
 ___   +     ___   +      ___   +      _____  +      ___
  |          |o           o|           -|-           |o
  |          |            |            |             |

 0|0         0|0          0|0          1|0           1|½
 ___         ___          ___          ___           ___
 0|0         0|0          0|0          0|0           0|½
```

A o echon is knocked to the +2 state in the formation of the quasi-particle. Straight-forward recombinations.

τ⁻ DECAY MODE	Fraction (Γ_i/Γ)	Scale factor/ Confidence level	p (MeV/c)
K⁻K⁺π⁻≥0neut. ν_τ	$(1.50 \pm 0.06) \times 10^{-3}$	S = 1.8	685
K⁻K⁺π⁻ν_τ	$(1.44 \pm 0.05) \times 10^{-3}$	S = 1.9	685

```
 τ⁻      g°     g°     q.p.      K⁻      K⁺     π⁻    ν_τ<18.2
1776..  70.0   70.0          493... 493... 139...  -i14.40
 -|      |      |     -o|o     o|      |o    |      -|
 ___  + _____+ _____ → _____ → ___  + ___ + ___ +  ___
  |     o|o    o|o      o|o     |      |     o|     |o
  |      |      |        |      |      |     |      |

 0|-½   1|1   -1|-1  1-1|-½    0|0    0|0   0|0    1|½
 ____   ___   _____  _____     ___    ___   ___    ___
 0|-½   0|1   0|-1   0|-½     -1|-1   0|0   0|0    0|½
  |_____|
    un-observed particles
```

63

A pair of o echons from a g^o graviton is knocked to the +2 state as the gravitons combine with the tauon to form a quasi-particle. Straight-forward recombinations.

		Scale factor/	p
τ **DECAY MODE**	Fraction (Γ_i/Γ)	Confidence level	(MeV/c)
$K^-K^+\pi^-\pi^0\nu_\tau$	$(6.1 \pm 2.5) \times 10^{-5}$	$S = 1.4$	618

```
   τ⁻              g°⁻             g°              q.p.
  1776..          70.0            70.0
   -|              |               |              -o|o
   ___            ___             ___             _____
   _|_     +    -o|o-    +       o|o      →      -o|o-    →
    |              |               |               |

  0|-½           3|2            -1|-1            3-1|½
  ____           ___            _____            _____
  0|-½           0|2             0|-1             0|½
                 |                        |
              un-observed particles
```

```
   K⁻             K⁺             π⁻             π⁰             ν_τ  <18.2
  493...         493...         139...         134...            -i14.40
   o|             |o             |              |               -|
   __            __             __             ___             ___
   _|_     +     _|_     +     o_|_     +     -_|-     +       _|o
    |             |             |              |               |

  0|0            0|0            0|0            1|0             1|½
  ___            ___            ___            ___             ___
  0|0            0|0            0|0            0|0             0|½
```

A negative and a positive o echon are knocked to the +2, as a g^{o-} and a g^o graviton combine with a tauon to form a quasi-particle. Then there are straight-forward recombinations.

$e^-e^-e^+\bar{\nu}_e\nu_\tau$ $(2.8 \pm 1.5)\text{x }10^{-5}$ 888

| τ^- | | g^{o-} | | g^- | | q.p. | |
| 1776.. | | 70.0 | | 70.0 | | | |

$$\begin{array}{c}{}-|\\ \hline \ \ |\ \ \\ \hline |\end{array} \quad + \quad \begin{array}{c}|\\ \hline -o\,|\,o- \\ \hline |\end{array} \quad + \quad \begin{array}{c}|\\ \hline -\,|\,- \\ \hline |\end{array} \quad \rightarrow \quad \begin{array}{c}-|\\ \hline o\,|\,o \\ \hline --\,|\,--\end{array} \quad \rightarrow$$

| $\dfrac{0\,|-\tfrac12}{0\,|-\tfrac12}$ | | $\dfrac{3\,|\,2}{0\,|\,2}$ | | $\dfrac{-1\,|\,-2}{0\,|\,-2}$ | | $\dfrac{3-1\,|-\tfrac12}{0\,|-\tfrac12}$ | |

$$|\rule{4cm}{0.4pt}|$$

un-observed particles

| e^- | | e^- | | e^+ | | $\bar{\nu}_e$ | | ν_τ <18.2 | |
| .510.. | | .510.. | | .510.. | | <.000010 | | -i14.40 | |

$$\begin{array}{c}|\\ \hline |\\ \hline -|\end{array} + \begin{array}{c}|\\ \hline |\\ \hline -|\end{array} + \begin{array}{c}|\\ \hline |\\ \hline |-\end{array} + \begin{array}{c}|\\ \hline o|\\ \hline |-\end{array} + \begin{array}{c}-|\\ \hline |o\\ \hline |\end{array}$$

| $\dfrac{0\,|-\tfrac12}{0\,|-\tfrac12}$ | | $\dfrac{0\,|-\tfrac12}{0\,|-\tfrac12}$ | | $\dfrac{0\,|-\tfrac12}{0\,|-\tfrac12}$ | | $\dfrac{1\,|\tfrac12}{0\,|\tfrac12}$ | | $\dfrac{1\,|\tfrac12}{0\,|\tfrac12}$ | |

A g^{o-} graviton and a g^- graviton combine with a tauon to form a quasi-particle. Four – echons are knocked to ground state. There are straight forward recombinations. The o echons go out in the neutrino and anti-neutrino.

Modes with five charged particles

$3h^-2h^+\geq0$ neutrals ν_τ $(1.02 \pm 0.04)\text{x }10^{-3}$ S = 1.1   794
(ex. $K^0_S \rightarrow \pi^-\pi^+$)
("5-prong")
$3h^-2h^+\nu_\tau(\text{ex.}K^0)$ $(8.39 \pm 0.35)\text{x }10^{-4}$ S = 1.1   794

τ⁻   g°   g°   g°   q.p.
1776.. 70.0   70.0   70.0

```
τ⁻          g°          g°          g°          q.p.
1776.. 70.0       70.0        70.0
 -|           |           |           |            -|
 _|    +    o|o   +    o|o   +    o|o   →   ooo|ooo     →
  |           |           |           |             |

0|-½       1|1        -1|-1       1|1        2-1|½
───        ───        ────        ───        ─────
0|-½       0|1        0|-1        0|1         0|½
   |_____|
            un-observed particles
```

```
π⁻          π⁻          π⁻          π⁺          π⁺          ντ <18.2
139...    139...     139...     139...     139...   -i14.40
 |           |           |           |           |            -|
o|    +    o|   +    o|   +    |o   +    |o   +    |o
 |           |           |           |           |             |

0|0        0|0        0|0        0|0        0|0        1|½
───        ───        ────       ───        ───        ───
0|0        1|1        -1|-1      0|0        0|0        0|½
```

Orbital spins of g° gravitons go out as positional-kinetic angular momentum for two of five pions and orbital spin of a tauon neutrino.

	Scale factor/	p
τ⁻ DECAY MODE	Fraction (Γ_i/Γ) Confidence level	(MeV/c)

$3h^-2h^+\pi^0\nu_\tau(ex.K^0)$ $(1.78 \pm 0.27)\text{x }10^{-4}$ 746

```
τ⁻          g°⁻         g°          g°          q.p.
1776.. 70.0       70.0        70.0
 -|           |           |           |            -|
 _|    +   -o|o-  +    o|o   +    o|o   →   -ooo|ooo-    →
  |           |           |           |             |

0|-½       3|2        -1|-1       -1|-1       3-2|-½
───        ───        ────        ────        ─────
0|-½       0|2        0|-1        0|-1         0|-½
   |_____|
        un-observed particles
```

π^-	π^-	π^-	π^+	π^+	π^0	ν_τ <18.2
139...	139...	139...	139...	139...	134...	-i14.40

```
  |         |         |         |         |         -|        -|
 o|    +   o|    +   o|    +    |o   +     |o   +   -|-   +    |o
  |         |         |         |         |          |         |


 0|0      0|0       0|0       0|0       0|0        1|0       1|½
 0|0      0|0       0|0       0|0       0|0       -1|-1      0|½
```

Straight-forward recombinations.

τ DECAY MODE	Fraction (Γ_i/Γ)	Scale factor/ Confidence level	p (MeV/c)

Miscellaneous other allowed modes

$(5\pi)^-\nu_\tau$	$(7.6 \pm 0.5) \times 10^{-3}$		800
$K^*(892)^- \geq 0h^0 \geq 0K^0_L \nu_\tau$	$(1.42 \pm 0.18)\%$	S = 1.4	665
$K^*(892)^- \nu_\tau$	$(1.20 \pm 0.07)\%$	S = 1.8	665

τ^-	g^-	g^0	q.p.	$K^*(892)^-$	ν_τ <18.2
1776..	>0i	>0i		891.6	-i14.40

```
 -|          |          |        -o|         o|         -|
  |    +    -|-   +    o|o   →   -|o-   →   -|-   +      |o
  |          |          |          |          |          |


 0|-½      -1|-2       1|1      1-1|-1½     1-1|-1       1|½
 0|-½       0|-2       0|1       0|-1½      -1|-2        0|½
           |                              |
       un-observed particles
```

The tauon knocks a o echon from the g^0 graviton to the +2 state as they and a g^- graviton combine to form a quasi-particle. Straight-forward recombinations occur. (For further information on the structure of the $K^*(892)^-$ particle, see Chapter 8, Chonomic Lists, Structure of Known Particles.)

τ DECAY MODE	Fraction (Γ_i/Γ)	Scale factor/ Confidence level	p (MeV/c)
$K^*(892)^0 K^- \geq 0$neutrals ν_τ	$(3.2 \pm 1.4) \times 10^{-3}$		542
$K^*(892)^0 K^- \nu_\tau$	$(2.1 \pm 0.4) \times 10^{-3}$		542

```
τ⁻        g°       g°       g⁻      q.p.      K*(892)⁰   K⁻    ν_τ <18.2
1776..   70.0     70.0     70.0              895.8      493... -i14.40
-|        |        |        |      -oo|        o|        o|      -|
_|   +   o|o  +   o|o  +   _|_  →  _-|oo-   →  _-|o-  +   _|  +   _|o
 |        |        |        |        |          |         |       |

0|-½     1|1      1|1     -1|-2  2-1|-½      1-1|-1      0|0     1|½
0|-½     0|1      0|1      0|-2    0|-½        0|-1      0|0     0|½
 |                                  |
       un-observed particles
```

A g⁻ graviton and two g° gravitons combine with a tauon to form a quasi-particle. Two negative o echonsares knocked up to the +2 state in the process. The tauon neutrino gets the +1 spin from one g° graviton. The K*(892)⁰ particle receives the +1 and -1 spins of the other two gravitons. There are straight-forward recombinations.

τ DECAY MODE	Fraction (Γ_i/Γ)	Scale factor/ Confidence level	p (MeV/c)
$\overline{K}*(892)^0\pi\geq 0$neutrals ν_τ	$(3.8 \pm 1.7)\times 10^{-3}$		655
$\overline{K}*(892)^0\pi^-\nu_\tau$	$(2.2 \pm 0.5)\times 10^{-3}$		655

```
τ⁻        g°       g°       g⁻      q.p.      K*(892)⁰   π⁻    ν̄_τ <18.2
1776..   70.0     70.0     70.0              895.5      139... -i14.40
-|        |        |        |      -o|         o|        |       -|
_|   +   o|o  +   o|o  +   _|_  →  _-o|oo-  →  _-|o-  +  o|  +    _|o
 |        |        |        |        |          |         |       |

0|-½     1|1      1|1     -1|-2  2-1|-½      1-1|-1      0|0     1|½
0|-½     0|1      0|1      0|-2    0|-½        0|-1      0|0     0|½
 |                                  |
       un-observed particles
```

This is very similar to the previous mode, except only one o echon is bumped to the +2 state.

τ DECAY MODE	Fraction (Γ_i/Γ)	Scale factor/ Confidence level	p (MeV/c)
$K_1(1270)^-\nu_\tau$	$(4.7 \pm 1.1)\times 10^{-3}$		433

```
  τ⁻       g⁺      g°         q.p.1      q.p.2    K₁(1270)⁻  ν_τ  <18.2
1776..   70.0    70.0                            1272       -i14.40

 -|        |       |         -o|        -o|        o|        -|
 _|   +  +|+    o|o    →   +|o+  →   -|o+  →   -|+   +  |o
  |        |       |          |         |         |         |

0|-½    1|2    -1|-1     2-1|1½      2|1½      1|1       1|½
0|-½    0|2     1|0       0|1½       0|1½      0|1       0|½
 |                 |
      un-observed
       particles
```

A g^+ and a g^o graviton are gravitationally attracted to a tauon, combine with it, and form a quasi-particle. The tauon echon knocks the negative o echon up to the +2 state. The -1 orbital spin in the incoming g^o graviton knocks over a + echon of the other graviton to a - echon. Thus the second quasi-particle is formed, which redivides into a $K_1(1270)^-$ particle and a tauon neutrino.

		Scale factor/	p
τ⁻ DECAY MODE	Fraction (Γ_i/Γ)	Confidence level	(MeV/c)
$K_1(1400)^-\nu_\tau$	(1.7 ± 2.6)x 10⁻³	S = 1.7	335

```
  τ⁻       g°         g°           q.p.     K₁(1400)⁻   ν_τ  <18.2
1776..   70.0       70.0                    1403        -i14.40

 -|        |          |          -o|         o|         -|
 _|   +  o|o   +  o|o    →   o|oo  →   o|o   +   |o
  |        |          |           |          |          |

0|-½   -1|-2      1|1      1-1|-1½     1-1|-1       1|½
0|-½    0|-2      0|1       0|-1½      -1|-2        0|½
 |                   |
      un-observed
       particles
```

Two g^o gravitons combine with a tauon to form a quasi-particle, which redivides into a $K_1(1400)^-$ and a tauon neutrino. Straight-forward recombinations.

τ **DECAY MODE**   Fraction (Γ_i/Γ)   Confidence level   (MeV/c)

$\eta\pi^-\pi^0\pi^0\,\nu_\tau$ (1.81 ± 0.31)x 10^{-4} 746

```
τ⁻        g⁺        g°        g⁻        g⁻        g⁻              q.p.
1776..  70.0      70.0      70.0      70.0      70.0
 -|        |         |         |         |         |               -|
 _|   +  +|+   +   o|o    +  -|-   +   -|-      +|+      →   --++o|o++--   →
  |        |         |         |         |         |               |

0|-½      1|2       1|1      -1|-2     -1|-2      1|2             1|½
0|-½      0|2       0|1       0|-2      0|-2      0|2             0|½
  |                                                    |
           un-observed particles
```

    η π^- π^0 $\pi^{\overline{0}}$ ν_τ <18.2

```
547.86    139...     134...     134...     -i14.40
  |          |          |          |          -|
-+|+-   +   o|    +   -|-    +   +|+     +    |o
  |          |          |          |           |

1-1|0       0|0        1|0       -1|0        1|½
 0|0        0|0        0|0        0|0         0|½
```

Straight forward recombinations.

Scale factor/   p

τ **DECAY MODE**   Fraction (Γ_i/Γ)   Confidence level   (MeV/c)

$\eta\pi^-\pi^0\pi^0\nu_\tau$ (1.81 ± 0.31)x 10^{-4} 746

```
τ⁻        g°⁻        g⁻        g⁻        g⁻            q.p.
1776..   70.0       70.0      70.0      70.0
 -|         |          |         |         |             -|
 _|   +  -o|o-    +  -|-   +   -|-   +   -|-    →   ----o|o----   →
  |         |          |         |         |              |

0|-½       3|2       -1|-2     -1|-2     -1|-2          0|-4½
0|-½       0|2        0|-2      0|-2      0|-2          0|-4½
   |         un-observed particles         |
```

```
   η          π⁻          π⁰           π⁰          ν_τ <18.2
547.86      139...      134...       134...        -i14.40
   |           |           |            |            -|
 __|__       __|__       __|__        __|__        __|__
 -+|+-   +   o|     +    -|-     +    -|-      +      |o
 __|__       __|__       __|__        __|__        __|__
   |           |           |            |            |

 1-1|0        0|0         1|0          1|0          1|½
 ____        ____        ____         ____         ____
 -1|-1       -1|-1       -1|-1        -1|-1        -1|-½
```

It takes a lot of positional-kinetic angular momentum to balance this mode. But there are straight forward recombinations.

| | | Scale factor/ | p |
τ DECAY MODE	Fraction (Γ_i/Γ)	Confidence level	(MeV/c)
ηK⁻ν_τ	(1.52 ± 0.08)x 10⁻⁴		719

```
  τ⁻        g°⁻        g⁺        q.p.          η          K⁻     ν_τ <18.2
1776..      70.0       70.0                  547.86      493..   -i14.40
 -|          |          |        -o|           |          o|      -|
 __|__      __|__      __|__      __|__        __|__      __|__   __|__
  |    +    -o|o-   +   +|+    →  -+|o+-   →   -+|+-   +    |        |o
 __|__      __|__      __|__      __|__        __|__      __|__   __|__
  |          |          |          |            |          |        |

 0|-½       3|2        1|2        4-1|2½       1-1|0       0|0      1|½
 ____       ____       ____       ____         ____        ____    ____
 0|-½       0|2        -1|1       0|2½         1|1         1|1     0|½
            |_____|
                    un-observed particles
```

A g°⁻ graviton and a g⁺ graviton combine with a tauon to form a quasi-particle, knocking a o echon to the 2 state.. There are straight-forward recombinations.

| | | Scale factor/ | p |
τ DECAY MODE	Fraction (Γ_i/Γ)	Confidence level	(MeV/c)
ηπ⁻π⁺π⁻ν_τ	(2.25 ± 0.13)x 10⁻⁴		743

```
  τ⁻         g⁺          g°          g°          g⁻              q.p.
1776..     70.0        70.0        70.0        70.0             
 -|          |           |           |          -|               -|
 ___   +    _|_    +    _|_    +    _|_    +    _|_    →    _____    →
  |         +|+         o|o         o|o         -|-         -+oo|oo+-
  |          |           |           |           |               |

 0|-½       1|2         1|1        -1|-1       -1|-2          2-2|-½
 ____       ___         ___        _____       _____          _____
 0|-½       0|2         0|1         0|-1        0|-2          0|-½
             |            un-observed particles              |
```

```
     η            π⁻           π⁺           π⁻          νₜ  <18.2
  547.86      139...       139...       139...       -i14.40
    |            |            |            |            -|
  _____   +    _|_    +     _|_    +     _|_    +     _|_
 -+|+-         o|           |o           o|           |o
    |            |            |            |            |

 1-1|0        0|0          0|0          0|0          1|½
 _____        ___          ____         ___          ___
  0|0         0|0         -1|-1         0|0          0|½
```

Four gravitons of various types combine with a tauon to form a quasi-particle. There are straight-forward recombinations.

τ DECAY MODE	Fraction (Γ_i/Γ)	Scale factor/ Confidence level	p (MeV/c)
$f_1(1285)\pi^-\nu_\tau$	$(3.9 \pm 0.5) \times 10^{-4}$		408
$f_1(1285)\pi^-\nu_\tau \to$ $\eta\pi^-\pi^+\pi^-\nu_\tau$	$(1.18 \pm 0.07) \times 10^{-4}$	S=1.3	--

```
   τ⁻           g°⁻          g°⁺                q.p.
1776..        70.0         70.0
  -|            |            |                  -|
 ___     +    _|_     +    _|_      →     _____      →
  |          -o|o-         +o|o+          -+oo|oo+-
  |            |            |                  |

 0|-½         3|2         -3|-2          3-3|-½
 ____         ___         _____          _____
 0|-½         0|2          0|-2          0|-½
              |        un-observed particles        |
```

72

f$_1$(1285)   π⁻   ν$_τ$ <18.2
  1281.9   139... -i14.40

```
     |              |           - |
 ____|____      ____|       _____|___
 -+o|o+-    +    o|     +       |o      →
     |              |             |

 2-1|1          0|0          1|½
 ───────        ─────       ──────
  0|1           -1|-1       -1|-½
```

```
   η          π⁻          π⁺          π⁻         ν$_τ$ <18.2
547.86      139...       139...      139...      -i14.40

    |           |           |           |           - |
 ___|___     ___|       ____|       ___|       _____|___
 -+|+-   +    o|    +      |o   +     o|    +      |o
    |           |           |           |           |

 1-1|0       0|0         0|0         0|0         1|½
 ─────       ────        ────        ────        ─────
  0|0        0|0         0|0         0|0         -1|-½
```

Two different gravitons combine with a tauon to form a quasi-particle, which first divides into a f$_1$(1285), a pi minus, and a tauon neutrino. The f$_1$(1285) further divides into an eta particle, a pi minus, and a pi plus.

τ⁻ DECAY MODE	Fraction (Γ$_i$/Γ)	Scale factor/ Confidence level	p (MeV/c)
h⁻ω≥0neutrals ν$_τ$	(2.41 ± 0.09)%	S = 1.2	708
h⁻ων$_τ$	(2.00 ± 0.08)%	S = 1.3	708

```
 τ⁻      g⁺      g°      q.p.1    q.p.2          π⁻      ω(782)   ν$_τ$<18.2
1776..  70.0    70.0                           139...   782.65   -i14.40
 -|       |       |       - |      - |           |        |        - |
 _|_  +  _|_  +  _|_  → __|___ → __|___ →  ____|    +  __|__  ____|___
  |      +|+      o|o     +o|o+     -o|o+       o|       -|+       |o
  |       |       |        |         |           |        |        |

 0|-½    1|2    -1|-1    2-1|1½    2|1½       0|0      1|1      1|½
 ────    ───    ─────    ──────    ─────      ────     ───      ───
 0|-½    0|2    1|0      0|1½      0|1½       0|0      0|1      0|½
            |un-observed particles|
```

Two gravitons combine with a tauon to form a quasi-particle. The minus orbital spin of the g^o graviton knocks over a + echon to a - echon, forming the second quasi-particle. Subsequently there are straight-forward recombinations, including the ω particle.

τ DECAY MODE	Fraction (Γ_i/Γ)	Scale factor/ Confidence level	p (MeV/c)
$h^-\omega\pi^0\nu_\tau$	$(4.1 \pm 0.4)\times 10^{-3}$		684
$h^-\omega 2\pi^0\nu_\tau$	$(1.4 \pm 0.5)\times 10^{-4}$		644

These two modes are similar to the above decay scheme, except there are additional one or two g^+ gravitons with -1 positional-kinetic angular momentum coming in, and one or two π^0s with +1 positional-kinetic angular momentum going out.

More Decay Modes are listed in J. Beringer *et al.* (Particle Data Group), PR **D86**, 010001 (2012) and 2013 partial update for the 2014 edition (URL:http://pdg.lbl.gov)

NEUTRINOS

. ν_e $J = \frac{1}{2}$

The following results are obtained using neutrinos associated with e^+ or e^-.

Mass m < 3 eV. Interpretation of tritium beta decay experiments is complicated by anomalies near the endpoint, and the limits are not without ambiguity.
Mean life/mass, $\tau/m_v > 7 \times 10^9$ s/eV (solar)
Mean life/mass, $\tau/m_{ve} > 300$ s/eV, CL = 90% (reactor)
Magnetic moment $\mu < 1.0 \times 10^{-10}$ μ_B, CL = 90%

Masses listed with the neutrinos below are from the calculations in Chapter 16.

$$\mathbf{\nu}_e \qquad\qquad\qquad\qquad \mathbf{\nu}_e$$
$$-i0.525 \text{ eV} \qquad\qquad -i0.525 \text{ eV}$$

or

depending on whether the electron neutrino has up or down spin. The down spin alternative will be taken as valid as well as the up spin alternative in all the data. For brevity, however, only the net positive spin alternatives are presented in most cases.

ν_μ $\qquad$ $J = \frac{1}{2}$

Mass $m < 0.19$ MeV, CL = 90%
Mean life/mass, $\tau/m_v > 15.4$ s/eV, CL = 90%
Magnetic moment $\mu < 6.8 \times 10^{-10}$ μ_B, CL = 90%

v_μ
<0.19 MeV
-i1.642 MeV

$$\frac{\;\;\mid\;\;}{-\mid\circ}$$
$$\mid$$

$$\frac{1\mid\frac{1}{2}}{\mid}$$

$\cdot\;\nu_\tau$ $\qquad\qquad\qquad$ $J = \frac{1}{2}$

Mass $m < 18.2$ MeV, CL = 95%
Magnetic moment $\mu < 3.9 \times 10^{-7}$ μ_B, CL = 90%
Electric dipole moment $d < 5.2 \times 10^{-17}$ e cm, CL = 95%

v_τ
<18.2 MeV
-i14.40 MeV

$$\frac{-\mid}{\mid\circ}$$
$$\mid$$

$$\frac{1\mid\frac{1}{2}}{\mid}$$

[1] J. Beringer *et al* (Particle Data Group) PR **D86**, 010001 (2012) and 2013 partial update for the 2014 edition (URL: http://pdg.lbl.gov).

[2] SUMMARY TABLES OF PARTICLE PROPERTIES, January 1 1998, Particle Data Group, *as quoted by CRC Handbook of Chemistry and Physics,* 80th Edition, David R. Lide, Ph.D, Editor in Chief (Boca Raton: CRC Press, 1999), pp. **11**-3 to **11**-5.

Chapter 8

Chonomic Lists

Structure of Known Particles

The quark and lepton model of particle physics divides charges in quarks to ±2e/3 and ±e/3. The electrino model of particle does not do that. Instead, it divides charges in electrinos to ±e, ±e/2, ±e/4, and ±e/8. The electrino model of particle physics does not hold that the quark and lepton model of particle physics is correct. Nevertheless, to facilitate cross referencing with the existing data, this volume will employ quark model titles and classifications in the subsequent classification of particles.

The chonomic structures and decay schemes contained in the following material are the author's, but the particle data come from [1]. "In this Summary Table:

"When a quantity has '(S = ...)' to its right, the error on the quantity has been enlarged by the 'scale factor' S, defined as

$$S = \sqrt{\chi^2 / (N - 1)},$$

where N is the number of measurements used in calculating the quantity. We do this when S > 1, which often indicates that the measurements are inconsistent. When S > 1.25, we also show in the Particle Listings an ideogram of the measurements. . . .

"A decay momentum p is given for each decay mode. For a 2-body decay, p is the momentum of each decay product in the rest frame of the decaying particle. For a 3-or-more-body decay, p is the largest momentum any of the products can have in this frame." [2]

To understand how to read and interpret chonomic structures and decay schemes, see Chapters 1-6.

For gravitons, the mass designations in the chonomic representations are 70.0 MeV. This is a derived result in this volume in Chapter 17. In Chapter 16, the masses of neutrinos are derived to be negative imaginary values, and positive imaginary values for anti-neutrinos. Experimental values of m^2 for neutrinos are also negative, indicating imaginary mass values for neutrinos (See note at end of this Chapter.)

Just as there is more than one way to skin a cat, there is more than one possible way to balance some decay schemes. For instance, the number of required incoming gravitons can vary, in some cases, depending on whether g^0, g^+, and g^- gravitons are employed, without incoming positional-kinetic angular momentum, or g^{o+} and g^{o-} gravitons are employed, with positional-kinetic angular momentum. Also unobserved incoming π^0s could be employed to balance decay schemes, though that is not necessary in this chapter. There are other flexible parameters in these decay schemes. Often the author just had to pick one example to solve the decay scheme. The following decay schemes illustrate a variety of ways to solve the problems. For a more advanced Summary Table, the various decay modes would have to be summed over the various possible pathways, to determine more fundamental quantities, such as populations of various graviton types.

In this chapter are listed state numbers with each chonomic structure. There is not room for them in Chapter 7. The state levels in this work, if not obvious to the investigator, may be obtained by looking up the particles in this chapter.

GAUGE AND HIGGS BOSONS

γ

$$I(J^{PC}) = 0,1(1^{--})$$

Mass $m < 1 \times 10^{-18}$ eV
Charge $q < 5 \times 10^{-35}$ e
Mean life τ = Stable

$$
\begin{array}{cc}
 & \gamma \\
 & 0 \\
2 & \bullet \mid \bullet \\
1 & \mid \\
0 & \mid \\
\end{array}
$$

$\pm 1 \mid \pm 1$

79

g
or gluon

$$I(J^P) = 0(1^-)$$

Mass m = 0
SU(3) color octet

No electrino formulation of gluons in chonomic system.

W

$J = 1$
Charge = ±1 e
Mass m = 80.385 ± 0.015 GeV
$m_Z - m_W$ = 10.4 ± 1.6 GeV
$m_{W+} - m_{W-}$ = -0.2 ± 0.6 GeV
Full width Γ = 2.085 ± 0.021 GeV
$<N_\pi^\pm>$ = 15.70 ± 0.35
$<N_K^\pm>$ = 2.20 ± 0.19
$<N_p>$ = 0.92 ± 0.14
$<N_{charged}>$ = 19.39 ± 0.08

```
                W
                ?
5            | o
4            |
3            |
2            |
1          + | +
0            |    ,

     1-1 | 1
            |
```

Z

$J = 1$
Charge = 0
Mass m = 91.1876 ± 0.0021 GeV
Full width Γ = 2.4952 ± 0.0023 GeV

```
          Z
          ?
5      + | +
4        |
3        |
2        |
1        |
0        |  ,

  ½-½ | 1
```

80

Higgs Bosons – H^0 and H$^{\pm}$

H^0 Mass m = 125.9 ± 0.4 GeV

High energy boson found. The author believes it will be found not to have Higgs Boson properties—that is, it will not define the masses of all particles. There is no formulation of Higgs Boson in chonomic system. The masses of particles are not defined by any one boson, but are self defined from first principles by each particle.

Unknown heavy or light bosons, searches for, are not covered in this appendix. This appendix covers known particles.

Citation: J. Beringer *et al.* (Particle Data Group), PR **D86**, 010001 (2012) and 2013 partial update for the 2014 edition (URL: http://pdg.lbl.gov)

LEPTONS

e

$J = \frac{1}{2}$

Mass m = (548.57990946 ± 0.00000022) x 10^{-6} u
Mass m = 0.510998928 ± 0.000000011 MeV
$|m_e^+ - m_e^-|/m < 8$ x 10^{-9}, CL = 90%

$$|q_{e^+} + q_{e^-}|/e < 4 \, x \, 10^{-8}$$

Magnetic moment anomaly
    (g-2)/2 = 1159.65218076 ± 0.00000027) x 10^{-6}

$$\left(g_{e^+} - g_{e^-}\right)/g_{average} = (-0.5 \pm 2.1) \, x \, 10^{-12}$$

Electric dipole moment d < 10.5 x 10^{-28} ecm, CL = 90%
Mean life $\tau > 4.6$ x 10^{26} yr, CL = 90%

e

Above measured mass is one input in model.

```
2      |
1      |
0     -|

     0|-½
       |
```

81

μ
$$J = \tfrac{1}{2}$$
Mass m = 105.6583715 ± 0.0000035 MeV
= 0.1134289267 ± 0.0000000029 u
Mean life τ = (2.1969811 ± 0.0000022) x 10^{-6} s
Magnetic moment anomaly (g-2)/2 = 11659209 ± 6) x 10^{-10}

```
             μ
105.671  336
2          |
1        - |
0          |


      0 | -½
        |
```

τ
$$J = \tfrac{1}{2}$$
Mass m = 1776.82 ± 0.16 MeV
```
         τ
1747.03 MeV
2      - |
1        |
0        |


    0 | -½
      |
```

NEUTRINOS

ν_e
$$J = \tfrac{1}{2}$$
Mass m <225 eV   95 CL%
```
       ν_e
-i0.525 eV
2        |
1        | o
0      - |


    1 | ½
      |
```

ν$_\mu$

$J = \frac{1}{2}$

Mass m <0.19 MeV 90 CL%

ν$_\mu$

-i1.642 MeV

```
2      |
1    - | o
0      |
```

$\underline{1 | \frac{1}{2}}$
        |

ν$_\tau$

$J = \frac{1}{2}$

Mass m <18.2MeV 95 CL%

ν$_\tau$

-i14.40 MeV

```
2    - |
1      | o
0      |
```

$\underline{1 | \frac{1}{2}}$
        |

QUARKS

No formulation of quarks in the electrino system.

ELECTRINOS

The chonomic system is for whole particles not electrinos, except for dots and anti-dots. For a treatise on the sizes and masses of electrinos, see *Electrino Physics*, Chapter 6, Section II E.

83

LIGHT UNFLAVORED MESONS
(S = C = B = 0)

$\pi^\pm$

$$I^G(J^P) = 1^-(0^-)$$

Mass m = 139.57018 ± 0.00035 MeV   (S = 1.2)

```
          π+
140.212  610  1  MeV
2          |
1          | o
0          |

         0 | 0
           |
```

π^0

$$I^G(J^{PC}) = 1^-(0^{-+})$$

Mass m = 134.9766 ± 0.0006 MeV   (S = 1.1)

```
          π0
                     MeV
2          |
1        - | -
0          |

         1 | 0
           |
```

η

$$I^G(J^{PC}) = 0^+(0^{-+})$$

Mass m = 547.862 ± 0.018 MeV

```
              η
      548.008 806  MeV
2          |
1        -+ | +-
0          |

    1½-1½ | 0
           |
```

---

f₀(500) $I^G(J^{PC}) = 0^+(0^{++})$
or σ

Mass m = (400-550) MeV

```
        f₀(500)
          ? MeV
2         |
1        o│o
0         |

   ½-½│0
      |
```

ρ(770)± $I^G(J^{PC}) = 1^+(1^{--})$
Mass m = 775.26 ± 0.25 MeV

```
          ρ+
          ?
2         |
1       -│o-
0         |

   1-1│-1
      |
```

ρ(770)⁰ $I^G(J^{PC}) = 1^+(1^{--})$
Mass m = 775.26 ± 0.25 MeV   (S = 1.8)

```
          ρ⁰

2         |
1      -o│o-
0         |

   2-1│-1
      |
```

ω(782) $I^G(J^{PC}) = 0^-(1^{--})$

Mass m = 782.65 ± 0.12 MeV   (S = 1.9)

```
       ω(782)
          ?
2         |
1       - | +
0         |

      1 | 1
        |
```

η'(958) $I^G(J^{PC}) = 0^+(0^{-+})$
Mass m = 957.78 ± 0.06 MeV

```
        η' (958)
           ?
2          |
1     - + o | o + -
0          |

  1½ - 1½ | 0
         |
```

$f_0(980)$ $I^G(J^{PC}) = 0^+(0^{++})$
Mass m = 990 ± 20 MeV

```
       f₀(980)
          ?
2       - | -
1       - | -
0         |

   2½ - ½ | 0
         |
```

86

$a_0(980)$ $\qquad$ $I^G(J^{PC}) = 1^-(0^{++})$

Mass m = 980 ± 20 MeV $\qquad$ (S = 1.5)

```
        a₀(980)
           ?
2          |
1       -+ | o+-
0          |

   1½-1½ | 0
          |
```

$\varphi(1020)$ $\qquad$ $I^G(J^{PC}) = 0^-(1^{--})$

Mass m = 1019.455 ± 0.020 MeV $\qquad$ (S =1.1)

```
         φ(1020)
            ?
2           |
1        -+o | o+-
0           |

    2-1 | 1
         |
```

$h_1(1170)$ $\qquad$ $I^G(J^{PC}) = 0^-(1^{+-})$

Mass m = 1170 ± 20 MeV

```
        h₁(1170)
           ?
2          |
1        -o | o-
0          |

   1½-1½ | -1
          |
```

87

$b_1(1235)$ $I^G(J^{PC}) = 1^+(1^{+-})$

Mass m = 1229.5 ± 3.2 MeV   (S = 1.6)

```
       b₁(1235)
          ?
2         |
1       - | o +
0         |

1½-½| 1
   |
```

$a_1(1260)$ $I^G(J^{PC}) = 1^-(1^{++})$
Mass m = 1230 ± 40 MeV

```
       a₁(1260)
          ?
2        | o
1       o | o
0         |

1½-½| 1
   |
```

$f_2(1270)$ $I^G(J^{PC}) = 0^+(2^{++})$
Mass m = 1275.1 ± 1.2 MeV   (S = 1.1)

```
       f₂(1270)
          ?
2       - | -
1      - o | o -
0         |

1½-1½| -2
    |
```

$f_1(1285)$ $\qquad$ $I^G(J^{PC}) = 0^+(1^{++})$

Mass $m = 1281.9 \pm 0.5$ MeV $\quad$ (S = 1.8)

```
      f1(1285)
         ?
2      - | -
1     -o | o-
0        |

     2-1 | -1
         |
```

$\eta(1295)$ $\qquad$ $I^G(J^{PC}) = 0^+(0^{-+})$
Mass $m = 1294 \pm 4$ MeV $\quad$ (S = 1.6)

```
       η(1295)
          ?
2       - | -
1     -+o | o+-
0         |

      2-1 | 0
          |
```

$\pi(1300)$ $\qquad$ $I^G(J^{PC}) = 1^-(0^{-+})$
Mass $m = 1300 \pm 100$ MeV

```
      π(1300)
         ?
2        | o
1       o | o
0        |

     1-1 | 0
         |
```

---

a$_2$(1320) $I^G(J^{PC}) = 1^-(2^{++})$

Mass m = 1318.3 ± 0.6 MeV   (S = 1.2)

```
      a₂(1320)
         ?
2       |o
1     -o|o-
0       |

     1-2|-2
        |
```

---

f$_0$(1370) $I^G(J^{PC}) = 0^+(0^{++})$

Mass m = 1200 to 1500 MeV

```
      f₀(1370)
         ?
2      -|-
1    -+o|o+-
0       |

     2-1|0
        |
```

π_1(1400) $I^G(J^{PC}) = 1^-(1^{-+})$

Mass m =1354 ± 25 MeV   (S = 1.8)

$$\pi_1\left(1400\right)$$

```
         ?
2       |o
1     -+o|o+-
0       |

     2-1|1
        |
```

η(1405)　　　　　　　　$I^G(J^{PC}) = 0^+(0^{-+})$
　　　　　　Mass m = 1408.8 ± 1.8 MeV　　(S = 2.1)

```
              η(1405 )
                 ?
      2       + | +
      1     -+o | o+-
      0         |

          1-2 | 0
              |
```

---

f₁(1420)　　　　　　　　$I^G(J^{PC}) = 0^+(1^{++})$
　　　　　　Mass m = 1426.4 ± 0.9 MeV　　(S = 1.1)

```
              f₁(1420)
                 ?
      2        - | -
      1     -+o | o+-
      0         |

        1½-1½ | -1
              |
```

ω(1420)　　　　　　　　$I^G(J^{PC}) = 0^-(1^{--})$

　　　　　Mass m = 1400 - 1450 MeV

```
              ω(1420)
                 ?
      2          |
      1      -o | o+
      0          |

          1-2 | -1
              |
```

91

$a_0(1450)$ $I^G(J^{PC}) = 1^-(0^{++})$

Mass m = 1474 ± 19 MeV

```
        a₀(1450)
           ?
2        | o
1       -+o | o+-
0         |

   1½-1½ | 0
         |
```

$\rho(1450)$ $I^G(J^{PC}) = 1^+(1^{--})$

Mass m = 1465 ± 25 MeV

```
        ρ(1450)
           ?
2        | o
1       -o | o-
0         |

   1½-1½ | -1
         |
```

$\eta(1475)$ $I^G(J^{PC}) = 0^+(0^{-+})$

Mass m = 1476 ± 4 MeV   (S = 1.3)

```
        η(1475)
           ?
2        | o
1       -+o | o+-
0         |

   1½-1½ | 0
         |
```

92

$f_0(1500)$ $I^G(J^{PC}) = 0^+(0^{++})$
Mass m = 1505 ± 6 MeV   (S = 1.3)

$f_0(1500)$
?

```
2        + | +
1       -o | o-
0          |

      1½-1½ | 0
          |
```

$f_2'(1525)$ $I^G(J^{PC}) = 0^+(2^{++})$
Mass m = 1525 ± 5 MeV

$f_2'(1525)$
?

```
2        - | -
1       -o | o-
0          |

      1½-1½ | -2
          |
```

$\pi_1(1600)$ $I^G(J^{PC}) = 1^-(1^{-+})$
Mass m = 1662^{+8}_{-9} MeV

$\pi_1(1600)$
?

```
2        - | -
1        - | o-
0          |

      2-1 | -1
          |
```

$\eta_2(1645)$ $I^G(J^{PC}) = 0^+(2^{-+})$
Mass m = 1617 ± 5 MeV

η(1645)
?
```
2      - | -
1      o | o
0        |
```

```
1-2 | -2
    |
```

$\omega(1650)$ $I^G(J^{PC}) = 0^-(1^{--})$

Mass m = 1670 ± 30 MeV

ω(1650)
?
```
2        |
1      -o | o+
0        |
```

```
1-2 | -1
    |
```

$\omega_3(1670)$ $I^G(J^{PC}) = 0^-(3^{--})$

Mass m = 1667 ± 4 MeV

$\omega_3(1670)$
?
```
2      o | o
1      - | +
0        |
```

```
-3 | -3
   |
```

$\pi_2(1670)$ $I^G(J^{PC}) = 1^-(2^{-+})$
Mass m = 1672.2 ± 3.0 MeV

```
         π₂(1670)
            ?
2        - | o -
1        o | o
0          |

        -1 | -2
           |
```

$\varphi(1680)$ $I^G(J^{PC}) = 0^-(1^{--})$

Mass m = 1680 ± 20 MeV

```
        φ(1680)
           ?
2        - | -
1      -+o | o+-
0          |

    1½-1½ | -1
          |
```

$\rho_3(1690)$ $I^G(J^{PC}) = 1^+(3^{--})$
Mass m = 1688.8 ± 2.1 MeV

```
        ρ₃(1690)
           ?
2        - | -
1        - | o -
0          |

      1-2 | -3
          |
```

$\rho(1700)$ $\qquad$ $I^G(J^{PC}) = 1^+(1^{--})$

Mass m = 1720 ± 20 MeV

```
        ρ(1700)
          ?
2        - | -
1         | o
0         |

   1-1 | -1
        |
```

$f_0(1710)$ $\qquad$ $I^G(J^{PC}) = 0^+(0^{++})$

Mass m = 1720 ± 6 MeV   (S = 1.6)

```
        f_j(1710)
          ?
2        o | o
1       -+ | +-
0         |

   1½-1½ | 0
        |
```

$\pi(1800)$ $\qquad$ $I^G(J^{PC}) = 1^-(0^{-+})$

Mass m = 1812 ± 12 MeV   (S = 2.3)

```
        π(1800)
          ?
2         | o
1       -+ | +-
0         |

   1½-1½ | 0
        |
```

96

$\varphi_3(1850)$ $I^G(J^{PC}) = 0^-(3^{--})$
Mass m = 1854 ± 7 MeV

```
           φ₃(1850)
              ?
 2         - | -
 1         o | o
 0           |

      ½-2½ | -3
           |
```

$\pi_2(1880)$ $I^G(J^{PC}) = 1^-(2^{-+})$
Mass m = 1895 ± 16 MeV

```
           φ₃(1880)
              ?
 2            |
 1         - | o-
 0         - | -

      1½-1½ | -2
           |
```

$f_2(1950)$ $I^G(J^{PC}) = 0^+(2^{++})$

Mass m = 1944 ± 12 MeV   (S = 1.5)

```
           f₂(1950)
              ?
 2         - | -
 1         -o | o-
 0           |

      1½-1½ | -2
           |
```

$f_2(2010)$ appears to be the same particle as $f_2(1950)$.

$f_2(2010)$ $I^G(J^{PC}) = 0^+(2^{++})$

        Mass m = 2011^{+60}_{-80} MeV

```
        f2(2010)
           ?
2       -o|o-
1        -|-
0         |

    1½-1½|-2
       |
```

$a_4(2040)$ $I^G(J^{PC}) = 1^-(4^{++})$
      Mass m = 1996^{+10}_{-9} MeV   (S = 1.1)

```
        a4(2040)
           ?
2       -o|o-        Appears to be the same as
1        -|-         f4(2050).
0         |

    ½-2½|-4
       |
```

$f_4(2050)$ $I^G(J^{PC}) = 0^+(4^{++})$
      Mass m = 2018 ± 11 MeV   (S = 2.1)

```
        f4(2050)
           ?
2       -o|o-        Appears to be the same as
1        -|-         a4(2040).
0         |

    ½-2½|-4
       |
```

$f_2(2300)$ $I^G(J^{PC}) = 0^+(2^{++})$

Mass $m = 2297 \pm 28$ MeV

```
        f₂(2300)
           ?
2      -o | o-              This appears to be the same as
1      -+ | +-              f₂(2340).
0         |

     1-2 | -2
          |
```

$f_2(2340)$ $I^G(J^{PC}) = 0^+(2^{++})$

Mass $m = 2339 \pm 60$ MeV

```
           ?
2      -o | o-
1      -+ | +-
0         |

     1-2 | -2
          |
```

This appears to be the identical particle as $f_2(2300)$. It has the same characteristics, the same decay products, and the mass of $f_2(2300)$ is within the error bar of the mass of $f_2(2340)$.

STRANGE MESONS
(S = ± 1, C = B = 0)

$K^\pm$ $I(J^P) = \frac{1}{2}(0^-)$

Mass $m = 493.677 \pm 0.016$ MeV   (S = 2.8)

```
          K⁺
   493.401 002 8 MeV
   2       | o
   1       |
   0       |

         0 | 0
           |
```

99

$. K_S^0$ $I(J^P) = \frac{1}{2}(0^-)$

Mass m = 497.614 ± 0.024 MeV   (S = 1.6)

K_S^0

```
        ?
2     - | -
1       |
0       |

      1 | 0
        |
```

$. K_L^0$ $I(J^P) = \frac{1}{2}(0^-)$

Mass m = 497.614 ± 0.024 MeV   (S = 1.6)

K_L^0

```
        ?
2     - |
1       | -
0       |

      1 | 0
        |
```

$K^*(892)^\pm$ $I(J^P) = \frac{1}{2}(1^-)$
.

Mass m = 891.66 ± 0.26 MeV

K* (892)$^+$

```
        ?
2       | o
1     - | -
0       |

    1-1 | -1
        |
```

100

K*(892)^0 $I(J^P) = \frac{1}{2}(1^-)$

Mass m = 895.81 ± 0.19 MeV   (S = 1.4)

```
      K*(892)⁰
         ?
2      __|o
1      o|
0       |

     -1|-1
        |
```

K$_1$(1270) $I(J^P) = \frac{1}{2}(1^+)$
Mass m = 1272 ± 7 MeV

```
      K₁(1270)
         ?
2      __|o
1     -o|+
0       |

     1-2|-1
        |
```

K$_1$(1400) $I(J^P) = \frac{1}{2}(1^+)$
Mass m = 1403 ± 7 MeV

```
      K₁(1400)
         ?
2      __|o
1     -o|-
0       |
   1½-1½|-1
        |
```

K*(1410) $I(J^P) = \frac{1}{2}(1^-)$

Mass m = 1414 ± 15 MeV   (S = 1.3)

```
        K*(1410)
           ?
2        o | o
1        - | -
0          |

  1½-1½ | -1
        |
```

K_0^* (1430) $I(J^P) = \frac{1}{2}(0^+)$

Mass m = 1425 ± 50 MeV

```
       K_0^* (1430)
           ?
2          | o
1        - o | -
0          |

   2-1 | 0
       |
```

$K_2^*(1430)^+$ $I(J^P) = \frac{1}{2}(2^+)$

Mass m = 1425.6 ± 1.5 MeV   (S = 1.1)

```
       K_2^*(1430)^+
           ?
2        - | o -
1        - | -
0          |

  1½-1½ | -2
        |
```

102

$\cdot K_2^*(1430)^0$ $\qquad$ $I(J^P) = \frac{1}{2}(2^+)$

Mass m = 1432.4 ± 1.3 MeV

$$K_2^*(1430)^0$$

```
            ?
2        - | o -
1        - o | -
0          |

     1½-1½ | -2
          |
```

$K^*(1680)$ $\qquad$ $I(J^P) = \frac{1}{2}(1^-)$

Mass m = 1717 ± 27 MeV $\qquad$ (S = 1.4)

$$K^*(1680)$$

```
            ?
2        o | o
1       - o | o -
0          |

     1½-1½ | -1
          |
```

$K_2(1770)$ $\qquad$ $I(J^P) = \frac{1}{2}(2^-)$

Mass m = 1773 ± 8 MeV

$$K_2(1770)$$

```
            ?
2        o | o
1       - o | o -
0          |

     1-2 | -2
        |
```

. $K_3^*(1780)$ $I(J^P) = \frac{1}{2}(3^-)$

Mass m = 1776 ± 7 MeV   (S = 1.1)

$$K_3^*(1780)$$

```
          ?
2        | o
1       -o | -
0         |

   ½-2½ | -3
      |
```

.

. $K_2(1820)$ $I(J^P) = \frac{1}{2}(2^-)$

Mass m = 1816 ± 13 MeV

$$K_2(1820)$$

```
          ?
2        | o
1       -o | -
0         |

   1-2 | -2
     |
```

. $K_4^*(2045)$ $I(J^P) = \frac{1}{2}(4^+)$

Mass m = 2045 ± 9 MeV   (S = 1.1)

$$K_4^*(2045)$$

```
          ?
2       -o | o-
1       -o | o-
0         |

   ½-2½ | -4
      |
```

104

CHARMED MESONS
(C = ± 1)

D$^{\pm}$ $I(J^P) = \frac{1}{2}(0^-)$
Mass m = 1869.62 ± 0.15 MeV   (S = 1.1)

$$D^+$$
```
1937.720 965 MeV
3        | o
2        |
1        |
0        |
```
```
0 | 0
  |
```

D^0 $I(J^P) = \frac{1}{2}(0^-)$
Mass m = 1864.86 ± 0.13 MeV   (S = 1.1)

$$D^0$$
```
3       - | -
2         |
1         |
0         |
```
```
1 | 0
  |
```

D*(2007)^0 $I(J^P) = \frac{1}{2}(1^-)$ I, J, P need confirmation
Mass m = 2006.99 ± 0.15 MeV
```
        D* (2007) 0
3       - | -
2         |
1       - | -
0         |
```
```
2 -1 | -1
   |
```

D*(2010)$^\pm$ $I(J^P) = \frac{1}{2}(1^-)$ I, J, P need confirmation.

Mass m = 2010.29 ± 0.13 MeV   (S = 1.1)

```
       D*(2010)+
          ?
3        _|o
2         |
1        -|-
0         |

        1-1|-1
          |
```

D$_1$(2420)^0 $I(J^P) = \frac{1}{2}(1^+)$ I, J, P need confirmation.

Mass m = 2421.4 ± 0.6 MeV   (S = 1.2)

```
          D1 (2420)0
             ?
3           -|-
2            |
1           o|o
0            |

        1½-1½|-1
            |
```

.$D_2^*(2460)^0$ $I(J^P) = \frac{1}{2}(2^+)$ $J^P = 2^+$ assignment strongly favored.
Mass m = 2462.6 ± 0.6 MeV   (S = 1.2)

$$D_2^*(2460)^0$$

```
             ?
3           -|-
2            |
1           -|-
0            |

        1½-1½|-2
            |
```

106

$.D_2^*(2460)^+$ $\quad$ $I(J^P) = \frac{1}{2}(2^+)$ $\quad$ $J^P = 2^+$ assignment strongly favored.

Mass m = 2464.3 ± 1.6 MeV $\quad$ (S = 1.7)

$$D_2^*(2460)^+$$

?

```
3        | o
2       -|-
1       -|-
0        |

    1½-1½ | -2
        |
```

CHARMED, STRANGE MESONS
(C = S = ± 1)

$D_S^\pm$

$.$was $F^\pm$

$\qquad$ $I(J^P) = 0(0^-)$

Mass m = 1968.50 ± 0.32 MeV $\quad$ (S = 1.3)

$$D_S^+$$

?

```
3        | o
1      o |
1       -|-
0        |

    2-1 | 0
        |
```

$.D_S^{*\pm}$ $\quad$ $I(J^P) = 0(?^?)$ $\quad$ J^P is natural, width and decay modes consistent with 1^-

Mass m = 2112.3 ± 0.5 MeV $\quad$ (S = 1.1)

$$D_S^{*+}$$

?

```
3        | o
2       -|-
1       -|-
0        |

    2-1 | -1
        |
```

$D_{s0}^{*}(2317)^{\pm}$ $I(J^P) = 0(0^+)$ J, P need confirmation.

Mass m = 2317.8 ± 0.6 MeV   (S = 1.1)

$$D_{s0}^{*}(2317)^{+}$$

?

```
2        o | o
2          |
1          |
0          |
```

½–½ | 0
|

$D_{s1}(2460)^{\pm}$ $I(J^P) = 0(1^+)$

Mass m = 2459.6 ± 0.6 MeV   (S = 1.1)

D$_{s1}$(2460)$^{\pm}$

?

```
3        | o
2        – | –
1          |
0          |
```

1–1 | –1
|

D$_{s1}$(2536)$^{\pm}$ $I(J^P) = 0(1^+)$ J, P need confirmation.

Mass m = 2535.12 ± 0.13 MeV

D$_{s1}$(2536)$^{+}$

?

```
3        o | o
2          | o
1        – | –
0          |
```

1½–1½ | –1
|

D*$_{s2}$(2573)   I(J^P) = 0($?^?$) J^P is natural, width and decay
     Mass m = 2571.9 ± 0.8 MeV   modes consistent with 2^+.
                  D$_{sJ}$(2573)$^+$
                     ?

3   o | o
2   o | o
1 |
0 |

    ½–2½ | –2
         |

BOTTOM MESONS
(B = ±1)

B$^\pm$ I(J^P) = ½(0^-)   I, J, P need confirmation. Quantum
      Mass m = 5279.26 ± 0.17 MeV   numbers shown are quark-
                  B$^+$ model predictions.
                   ?

4 | o
3 |
2 |
1 |
0 |

      0 | 0
        |

B^0 I(J^P) = ½(0^-) I, J, P need confirmation. Quantum numbers shown are
      Mass m = 5279.58 ± 0.17 MeV   quark-model predictions.
                  B^0
                   ?

4   – | –
3 |
2 |
1 |
0 |

     1 | 0
       |

109

B* $I(J^P) = \frac{1}{2}(1^-)$ I, J, P need confirmation. Quantum numbers shown are
Mass m = 5325.1\2 ± 0.4 MeV   quark-model predictions.

```
                    B*
                    ?
4                 | o
3                 |
2                 |
1               - | -
0                 |


        1 - 1 | - 1
            |
```

$B_1(5721)^0$ $I(J^P) = \frac{1}{2}(1^+)$ I, J, P need confirmation.
Mass m = 5723.5 ± 2.0 MeV   (S = 1.1)

$$B_1(5721)^0$$

```
                    ?
4               o | o
3                 |
2                 |
1               - | -
0                 |


      1½ - 1½ | - 1
            |
```

$B_2^*(5747)^0$ $I(J^P) = 1/2(2+)$ I, J, P need confirmation.
Mass m = 5743.9 ± 5 MeV   (S = 2.9)

$$B_2^*(5747)^0$$

```
                    ?
4               o | o
3                 |
2               o | o
1               - | -
0                 |


          3 | 2
            |
```

$.B_S^0$ $I(J^P) = 0(0^-)$ I, J, P need confirmation. Quantum numbers
Mass m = 5366.77 ± 0.24 MeV   shown are quark-model

$$B_S^0$$
?

4 | o
3 |
2   o |
1 |
0 |

½–½ | 0
|

predictions.

B_s^* $I(J^P) = 0(1^-)$ I, J, P need confirmation. Quantum numbers

Mass m = 5415.4 ± 2.4 MeV   shown are quark-model

$$B_S^*$$
?

4   – | –
3 |
2   – | –
1 |
0 |

3 | 1
|

predictions.

$B_{s1}(5830)^0$ $I(J^P) = ½(1^+)$ I, J, P need confirmation.
Mass m = 5828.7 ± 0.4 MeV

$$B_{s1}(5830)^0$$
?

4   o | o
3 |
2   – | –
1 |
0 |

1½–1½ | –1
|

$B_{s2}^*(5840)^0$ $\qquad$ $I(J^P) = \frac{1}{2}(2^+)$ $\quad$ I, J, P need confirmation.

$\qquad\qquad$ Mass m = 5839.96 ± 0.20 MeV

$$B_{s1}(5840)^0$$

?

4	o | o
3	|
2	− | −
1	− | −
0	|

$1\frac{1}{2}$−$1\frac{1}{2}$ \| −2

\|

BOTTOM, CHARMED MESONS
(B = C = ± 1)

$B_c^\pm$ $\qquad$ $I(J^P) = 0(0^-)$ $\quad$ I, J, P need confirmation. Quantum numbers shown

$\qquad\qquad$ Mass m = 6.2745 ± 0.0018 MeV $\qquad$ are quark-model predictions.

$$B_c^+$$

?

4	| o
3	− | −
2	|
1	|
0	|

$1\frac{1}{2}$−$\frac{1}{2}$ \| 0

\|

$$\bar{c}c \text{ MESONS}$$

$\eta_c(1S)$ $\qquad\qquad$ $I^G(J^{PC}) = 0^+(0^{-+})$

$\qquad$ Mass m = 2983.7 ± 0.7 MeV $\qquad$ (S = 1.4)

η_c(1S)

?

3	o | o
2	| o
1	− + | + −
0	|

$1\frac{1}{2}$−$1\frac{1}{2}$ \| 0

\|

J/ψ(1S) $I^G(J^{PC}) = 0^-(1^{--})$
Mass m = 3096.916 ± 0.011 MeV

```
        J/ψ(1S)
           ?
3       o | o
2         | o
1       - | +
0         |

    1-2 | -1
         |
```

χc0(1P) $I^G(J^{PC}) = 0^+(0^{++})$
Mass m = 3414.75 ± 0.31 MeV

```
        χc0(1P)
           ?
3       o | o
2       o | o
1         |
0         |

    1½-1½ | 0
          |
```

χc1(1P) $I^G(J^{PC}) = 0^+(1^{++})$

Mass m = 3510.66 ± 0.07 MeV

```
        χc1(1P)
           ?
3       o | o
2       o | o
1       - | -
0         |

    1½-1½ | -1
           |
```

113

$h_c(1P)$

$$I^G(J^{PC}) = ?^?(1^{+-})$$

Mass m = 3525.38 ± 0.11 MeV

$$h_c(1P)$$

```
           ?
3       o | o
2      -o | o-
1        |
0        |

    1½-1½| -1
         |
```

$\chi_{c2}(1P)$

$$I^G(J^{PC}) = 0^+(2^{++})$$

Mass m = 3556.20 ± 0.09 MeV

$$\chi_{c2}(1P)$$

```
           ?
3       o | o
2      -o | o-
1       - | -
0        |

    1½-1½| -2
         |
```

$\eta_c(2S)$

$$I^G(J^{PC}) = 0^+(0^{-+})$$

Mass m = 3639.4 ± 1.3 MeV   (S = 1.2)

$$\eta_c(2S)$$

```
           ?
3       o | o
2       o | o
1      -+ | +-
0        |

    1½-1½| 0
         |
```

$\psi(2S)$ $I^G(J^{PC}) = 0^-(1^{--})$
Mass m = 3686.109 ± 0.014 MeV   (S = 1.6)

```
        ψ(2S)
          ?
3       o | o
2       o | o
1       - | +
0         |

      1-2 | -1
          |
```

$\psi(3770)$ $I^G(J^{PC}) = 0^-(1^{--})$

Mass m = 3773.15 ± 0.33 MeV
```
       ψ(3770)
          ?
3       o | o
2       o | o
1       + | -
0         |

      1-2 | -1
          |
```

$\chi(3872)$ $I^G(J^{PC}) = 0^?(?^{?+})$
Mass m = 3871.68 ± 0.17 MeV

$\chi(3872)$
```
          ?
3       o | o
2       o | o
1       - | +
0         |

       -3 | -3
          |
```

115

$\psi(4040)$ $\qquad$ $I^G(J^{PC}) = 0^-(1^{--})$
Mass m = 4039 ± 1 MeV

```
        ψ(4040)
           ?
  3      o | o
  2      o | o
  1     -+ | +-
  0        |

       1-2 | -1
          |
```

$\psi(4160)$ $\qquad$ $I^G(J^{PC}) = 0^-(1^{--})$
Mass m = 4153 ± 3 MeV

```
        ψ(4160)
           ?
  3       o | o
  2      -o | o-
  1      -+ | +-
  0         |

     1½-1½ | -1
          |
```

$\chi(4260)$ $\qquad$ $I^G(J^{PC}) = ?^?(1^{--})$

Mass m = 4250 ± 9 MeV $\qquad$ (S = 1.6)

```
        χ(4260)
           ?
  3       o | o
  2     -+o | o+-
  1      -+ | +-
  0         |

     1½-1½ | 1
          |
```

116

ψ(4415)　　　　　$I^G(J^{PC}) = 0^-(1^{--})$

Mass m = 4421 ± 4 MeV
```
           ψ(4415)
             ?
3        o | o
2      -+o | o+-
1      -+o | o+-
0          |

     1½-1½ | -1
             |
```

$b\bar{b}$ MESONS

Y(1S)　　　　　　$I^G(J^{PC}) = 0^-(1^{--})$
Mass m = 9460.30 ± 0.26 MeV   (S = 3.3)
```
            Y(1S)
              ?
4        o | o
3          |
2          | o
1          |
0          |

      ½-1½ | -1
             |
```

χ_{b0}(1P)　　　　　　$I^G(J^{PC}) = 0^+(0^{++})$ J needs confirmation.
Mass m = 9859.44 ± 0.42 MeV
```
          χ_{b0}(1P)
              ?
4        o | o
3          |
2          |
1        o | o
0          |

     1½-1½ | 0
             |
```

$\chi_{b1}(1P)$ $I^G(J^{PC}) = 0^+(1^{++})$ J needs confirmation.

Mass m = 9892.78 ± 0.26 MeV

$\chi_{b1}(1P)$

?

```
4        o | o
3          |
2        o | o
1        - | -
0          |


  1½–1½ | –1
       |
```

$\chi_{b2}(1P)$ $I^G(J^{PC}) = 0^+(2^{++})$ J needs confirmation.

Mass m = 9912.21 ± 0.26 MeV

$\chi_{b2}(1P)$

?

```
4        o | o
3          |
2        o | o
1        - | -
0          |


  1–2 | –2
      |
```

Y(2S) $I^G(J^{PC}) = 0^-(1^{--})$

Mass m = 10.02326 ± 0.00031 GeV

Y(2S)

?

```
4        o | o
3          |
2        o | o
1          |
0          |


  1–2 | –1
      |
```

118

$\chi_{b0}(2P)$ $I^G(J^{PC}) = 0^+(0^{++})$ J needs confirmation.

Mass m = 10.2325 ± 0.0004 GeV

$\chi_{b0}(2P)$

?

```
4        o | o
3          |
2        o | o
1        -+ | +-
0          |

    1½-1½ | 0
          |
```

$\chi_{b1}(2P)$ $I^G(J^{PC}) = 0^+(1^{++})$ J needs confirmation.

Mass m = 10.25546 ± 0.00022 GeV

$\chi_{b1}(2P)$

?

```
4        o | o
3          |
2       -o | o-
1       -+ | +-
0          |

    1½-1½ | -1
          |
```

$\chi_{b2}(2P)$ $I^G(J^{PC}) = 0^+(2^{++})$ J needs confirmation.

Mass m = 10.26865 ± 0.00022 GeV

$\chi_{b2}(2P)$

?

```
4        o | o
3          |
2       -o | o-
1        - | -
0          |

    1½-1½ | -2
          |
```

119

Y(3S) $\qquad$ $I^G(J^{PC}) = 0^-(1^{--})$

Mass m = 10.3552 ± 0.0005 GeV

Y(3S)

?

```
4    o | o
3      |
2    -o | o-
1      |
0      |
```

$1\frac{1}{2}-1\frac{1}{2} | -1$
$\qquad\quad |$

Y(4S) $\qquad$ $I^G(J^{PC}) = 0^-(1^{--})$
or Y(10580)

Mass m = 10.5794 ± 0.0012 GeV

Y(4S)

?

```
4    o | o
3      |
2    -o | o-
1    - | -
0      |
```

$2-1 | -1$
$\quad\; |$

Y(10860) $\qquad$ $I^G(J^{PC}) = 0^-(1^{--})$

Mass m = 10.876 ± 0.011 GeV   (S = 1.1)

Y(10860)

?

```
4    o | o
3      |
2    -o | o-
1    - | +
0      |
```

$1\frac{1}{2}-1\frac{1}{2} | -1$
$\qquad\quad |$

120

Y (11020) $I^G(J^{PC}) = 0^-(1^{--})$
Mass m = 11.020 ± 0.008 GeV
Y (11020)
?

```
4      o | o
3      − | −
2     −o | o−
1        |
0        |
```

2−1 | −1
|

N BARYONS
(S = 0, I = ½)

p $I(J^P) = \frac{1}{2}(\frac{1}{2}^+)$

Mass m = 938.272046 ± 0.000021 MeV
p
?

```
2      | ●
1      | o
0     +|
```

1−1 | ½
|

---

n $I(J^P) = \frac{1}{2}(\frac{1}{2}^+)$
Mass m = 939.56536 ± 0.00008 MeV
n
934.209 320 9 MeV

```
2      | ●
1     +|
0      |
```

−1 | −½
|

N(1440) ½ $^+$ I(J^P) = ½(½$^+$)

Breit-Wigner mass = 1420 to 1470 ($\approx$ 1440) MeV

```
         N(1440)1/2⁺
             ?
   2       | ●
   1      −+ | o+
   0        |

      1½−1½|½
          |
```

N(1520) 3/2$^-$ I(J^P) = ½(3/2$^-$)

Breit-Wigner mass = 1515 to 1525 ($\approx$ 1520) MeV

```
         N(1520)3/2⁻
             ?
   2        | ●
   1      −+ | o−
   0        |

     −1½ | −1½
         |
```

N(1535) ½$^-$ I(J^P) = ½(½$^-$)

Breit-Wigner mass = 1525 to 1545 ($\approx$ 1535) MeV

```
         N(1535)1/2⁻
             ?
   2        | ●
   1       − | o−
   0       +|

    1½−1½ | −½
         |
```

N(1650) $\frac{1}{2}^-$ $I(J^P) = \frac{1}{2}(\frac{1}{2}^-)$

Breit-Wigner mass = 1645 to 1670 ($\approx$ 1655) MeV

```
        N(1650)1/2⁻
            ?
  2     + | ●
  1     − | o−
  0       |

  1½−1½ | −½
        |
```

N(1675) $\frac{5}{2}^-$ $I(J^P) = \frac{1}{2}(5/2^-)$

Breit-Wigner mass = 1670 to 1680 ($\approx$ 1675) MeV

```
         N(1675)5/2⁻
             ?
  2      + | ●
  1      + | +
  0        |

   2−1 | 2½
        |
```

N(1680) $5/2^+$ $I(J^P) = \frac{1}{2}(5/2^+)$

Breit-Wigner mass = 1680 to 1690 ($\approx$ 1685) MeV

```
         N(1680)5/2⁺
             ?
  2      + | ●
  1      + | o+
  0        |

   2−1 | 2½
        |
```

N(1700) 3/2⁻   I(J^P) = ½(3/2⁻)

Breit-Wigner mass = 1650 to 1750 (≈ 1700) MeV

```
        N(1700)3/2⁻
             ?
2      + | ●
1      + | o+
0        |

    1½−1½ | 1½
          |
```

N(1710) ½⁺   I(J^P) = ½(½⁺)

Breit-Wigner mass = 1680 to 1740 (≈ 1710) MeV

```
        N(1710)1/2⁺
             ?
2      + | ●
1      + | o+
0        |

     1−2 | ½
```

```
          |
```

N(1720) 3/2⁺   I(J^P) = ½(3/2⁺)

Breit-Wigner mass = 1700 to 1750 (≈ 1720) MeV

```
        N(1720)3/2⁺
             ?
2      + | ●
1     −+ | o+−
0        |

     2−1 | 1½
          |
```

N(2190) 7/2⁻ → $N(2190) 7/2^-$ $I(J^P) = \frac{1}{2}(7/2^-)$

Breit-Wigner mass = 2100 to 2200 ($\approx$ 2190) MeV

```
        N(2190) 7/2⁻
            ?
  2       + | ●
  1       + | o +
  0         |

       2½ –½ | 3½
            |
```

N(2220) 9/2⁺ $I(J^P) = \frac{1}{2}(9/2^+)$

Breit-Wigner mass = 2200 to 2300 ($\approx$ 2250) MeV

```
        N(2220) 9/2⁺
            ?
  2       + | ●    This appears to be the same
  1       + | o +  particle as N(2250)G₁₉.
  0         |

        3 | 4½
          |
```

N(2250) 9/2⁻ $I(J^P) = \frac{1}{2}(9/2^-)$

Breit-Wigner mass = 2200 to 2350 ($\approx$ 2275) MeV

```
        N(2250) 9/2⁻
            ?
  2       + | ●    This appears to be the same
  1       + | o +  particle as N(2220)H₁₉.
  0         |

        3 | 4½
          |
```

N(2600) 11/2⁻ $I(J^P) = \frac{1}{2}(11/2^-)$

Breit-Wigner mass = 2550 to 2750 ($\approx$ 2600) MeV

```
        N(2600)11/2⁻
             ?
3         + | +
2         + | •
1         + | ○ +
0           |

         3 | 5½
           |
```

Δ BARYONS
(S = 0, I = 3/2)

Δ(1232) 3/2⁺ $I(J^P) = 3/2(3/2^+)$

Breit-Wigner mass (mixed charges) = 1230 to 1234 ($\approx$ 1232) MeV

```
       Δ(1232) 3/2⁺
            ?
2          | •
1         ○ | ○
0         + |

       2-1 | 1½
           |
```

Δ(1600) 3/2⁺ $I(J^P) = 3/2(3/2^+)$

Breit-Wigner mass = 1500 to 1700 ($\approx$ 1600) MeV

```
       Δ(1600) 3/2⁺
            ?
2          | •
1        + ○ | ○
0          |

       2-1 | 1½
           |
```

126

$\Delta(1620)$ ½⁻ $I(J^P) = 3/2(1/2^-)$

Breit-Wigner mass = 1600 to 1660 (≈ 1630) MeV

```
              Δ(1620)1/2⁻
                  ?
    2          |●
    1        −o|o
    0          |

      1½−1½|−½
           |
```

$\Delta(1700)$ 3/2⁻ $I(J^P) = 3/2(3/2^-)$

Breit-Wigner mass = 1670 to 1750 (≈ 1700) MeV

```
              Δ(1700)3/2⁻
                  ?
    2         −|●
    1        −o|o−
    0          |

      1½−1½|−1½
           |
```

$\Delta(1905)$ 5/2⁺ $I(J^P) = 3/2(5/2^+)$

Breit-Wigner mass = 1855 to 1910 (≈ 1880) MeV

```
              Δ(1905)5/2⁺
                  ?
    2         +|●
    1        +o|o+
    0          |

      2−1|2½
          |
```

$\Delta(1910)$ ½$^+$ $I(J^P) = 3/2(1/2^+)$

Breit-Wigner mass = 1860 to 1910 ($\approx$ 1890) MeV

```
            Δ(1910) P₃₁
                 ?
  2           + | ●
  1           ○ | ○
  0             |

      1½−1½ | ½
            |
```

$\Delta(1920)$ 3/2$^+$ $I(J^P) = 3/2(3/2^+)$

Breit-Wigner mass = 1900 to 1970 ($\approx$ 1920) MeV

```
           Δ(1920) 3/2⁺
                 ?
  2           + | ●
  1           +○ | ○+
  0              |

      1½−1½ | 1½
            |
```

$\Delta(1930)$ 5/2$^-$ $I(J^P) = 3/2(5/2^-)$

Breit-Wigner mass = 1900 to 2000 ($\approx$ 1950) MeV

```
           Δ(1930) 5/2⁻
                 ?
  2             | ●
  1           −○ | ○
  0             |

      ½−2½ | −2½
            |
```

128

$\Delta(1950)\ 7/2^+$ $I(J^P) = 3/2(7/2^+)$

Breit-Wigner mass = 1915 to 1950 ($\approx$ 1930) MeV

$\Delta(1950)\ 7/2^+$

```
            ?
2          | •
1        +○ | ○
0          |

        3 | 3½
          |
```

$\Delta(2420)\ 11/2^+$ $I(J^P) = 3/2(11/2^+)$

Breit-Wigner mass = 2300 to 2500 ($\approx$ 2420) MeV

$\Delta(2420)\ 11/2^+$

```
            ?
3        + | +
2        + | •
1       +○ | ○+
0          |

        3 | 5½
          |
```

Λ BARYONS
(S = -1, I = 0)

Λ $I(J^P) = 0(\frac{1}{2}^+)$

Mass m = 1115.683 $\pm$ 0.006 MeV

```
            Λ
   1038.603 161 MeV
2          | •
1        + |
0          |

       -1 | -½
          |
```

$\Lambda(1405)\ \frac{1}{2}^-$ $\qquad$ $I(J^P) = 0(1/2^-)$

Mass m = 1405.1 ± 1.3 MeV

$$\Lambda(1405)\ \frac{1}{2}^-$$

```
              ?
2        _|●
1        −|
0         |

  ⅛−⅛ | −½
      |
```

$\Lambda(1520)\ 3/2^-$ $\qquad$ $I(J^P) = 0(3/2^-)$

Mass m = 1519.5 ± 1.0 MeV

$$\Lambda(1520)\ 3/2^-$$

```
              ?
2        _|●
1        −|
0         |

  −1 | −1½
     |
```

$\Lambda(1600)\ \frac{1}{2}^+$ $\qquad$ $I(J^P) = 0(1/2^+)$

Mass m = 1560 to 1700 (≈ 1600) MeV

$$\Lambda(1600)\ \frac{1}{2}^+$$

```
              ?
2        _|●
1        +|
0         |

  ½−½ | ½
     |
```

$\Lambda(1670)$ ½⁻ $I(J^P) = 0(1/2^-)$

Mass m = 1660 to 1680 (≈ 1670) MeV

$\Lambda(1670)$ ½⁻
?

```
2      – | ●
1      – | –
0        |
```

$$\frac{2-1 \mid -\frac{1}{2}}{|}$$

$\Lambda(1690)$ 3/2⁻ $I(J^P) = 0(3/2^-)$

Mass m = 1685 to 1695 (≈ 1690) MeV

$\Lambda(1690)$ 3/2⁻
?

```
2      – | ●
1      – | –
0        |
```

$$\frac{1\frac{1}{2}-1\frac{1}{2} \mid -1\frac{1}{2}}{|}$$

$\Lambda(1800)$ ½⁻ $I(J^P) = 0(1/2^-)$

Mass m = 1720 to 1850 (≈ 1800) MeV

$\Lambda(1800)$ ½⁻
?

```
2        | ●
1      –o | o
0        |
```

$$\frac{1\frac{1}{2}-1\frac{1}{2} \mid -\frac{1}{2}}{|}$$

131

$\Lambda(1810)$ ½$^+$ $I(J^P) = 0(1/2^+)$

Mass m = 1750 to 1850 ($\approx$ 1810) MeV

$\Lambda(1810)$ ½$^+$
?
```
2        | ●
1      −+o | o+
0        |
```

```
1½−1½ | ½
      |
```

$\Lambda(1820)$ 5/2$^+$ $I(J^P) = 0(5/2^+)$

Mass m = 1815 to 1825 ($\approx$ 1820) MeV

$\Lambda(1820)$ 5/2$^+$
?
```
2       + | ●
1      +o | o+
0        |
```

```
2−1 | 2½
   |
```

$\Lambda(1830)$ 5/2$^-$ $I(J^P) = 0(5/2^-)$

Mass m = 1810 to 1830 ($\approx$ 1820) MeV

$\Lambda(1830)$ 5/2$^-$
?
```
2       − | ●
1      −o | o−
0        |
```

```
1−2 | −2½
   |
```

$\Lambda(1890)\,3/2^+$ $\qquad$ $I(J^P) = 0(3/2^+)$

Mass m = 1850 to 1910 ($\approx$ 1890) MeV

$$\Lambda(1890)\,3/2^+$$

```
           ?
2      + | •
1      +o | o+
0         |

   1½–1½ | 1½
         |
```

$\Lambda(2100)\,7/2^-$ $\qquad$ $I(J^P) = 0(7/2^-)$

Mass m = 2090 to 2110 ($\approx$ 2100) MeV

$$\Lambda(2100)\,7/2^-$$

```
          ?
2        | •
1      –o | o
0        |

    –3 | –3½
       |
```

$\Lambda(2110)\,5/2^+$ $\qquad$ $I(J^P) = 0(5/2^+)$

Mass m = 2090 to 2140 ($\approx$ 2110) MeV

$$\Lambda(2110)\,5/2^+$$

```
          ?
3      + | +
2      + | •
1      o | o
0        |

    2–1 | 2½
        |
```

$\Lambda(2350)\ 9/2^+$ $I(J^P) = 0(9/2^+)$

Mass m = 2340 to 2370 ($\approx$ 2350) MeV

$$\Lambda(2350)\ 9/2^+$$

```
                ?
3          + | +
2          + | ●
1          +○ | ○+
0            |

        2½−½ | 4½
             |
```

Σ BARYONS
(S = -1, I = 1)

Σ^+ $I(J^P) = 1(\frac{1}{2}^+)$

Mass m = 1189.37 $\pm$ 0.07 MeV   (S = 2.2)

$$\Sigma^+$$

```
             ?
2       ○ | ●+
1         |
0         |

      1−1 | ½
          |
```

Σ^0 $I(J^P) = 1(\frac{1}{2}^+)$

Mass m = 1192.642 $\pm$ 0.024 MeV

$$\Sigma^0$$

```
1169.415  992 MeV
2       + | ●
1         |
0         |

      ½−½ | ½
          |
```

134

Σ^- $I(J^P) = 1(1/2^+)$

Mass m = 1197.449 ± 0.030 MeV   (S = 1.2)

```
              Σ⁻
              ?
   2       +○ | ●
   1          |
   0          |

        1−1 | ½
            |
```

$\Sigma(1385)^+ 3/2^+$ $I(J^P) = 1(3/2^+)$

Mass m = 1382.80 ± 0.35 MeV   (S = 2.0)

```
           Σ(1385)⁺ 3/2⁺
                ?
   2        ○ | ●+
   1        + | +
   0          |

     1½−1½ | 1½
           |
```

$\Sigma(1385)^0 3/2^+$ $I(J^P) = 1(3/2^+)$

Mass m = 1383.70 ± 1.0 MeV   (S = 1.4)

```
           Σ(1385)⁰ 3/2⁺
                ?
   2        + | ●
   1        + | +
   0          |

     1½−1½ | 1½
           |
```

$\dot{\Sigma}(1385)^- \ 3/2^+$ $I(J^P) = 1(3/2^+)$

Mass m = 1387.2 ± 0.5 MeV   (S = 2.2)

$\Sigma(1385)^- \ 3/2^+$

?

2	+○ | ●
1	+ | +
0	|

1½–1½ \| 1½

\|

$\Sigma(1660) \ \frac{1}{2}^+$ $I(J^P) = 1(1/2^+)$

Mass m = 1630 to 1690 (≈ 1660) MeV

$\Sigma(1660) \ P \ \frac{1}{2}^+$

?

2	+ | ●
1	○ | ○
0	|

1½–1½ \| ½

\|

$\dot{\Sigma}(1670) \ 3/2^-$ $I(J^P) = 1(3/2^-)$

Mass m = 1665 to 1685 (≈ 1670) MeV

$\Sigma(1670) \ 3/2^-$

?

2	+ | ●
1	○ | ○
0	|

2–1 \| 1½

\|

136

$\Sigma(1750)$ ½⁻ $\qquad$ $I(J^P) = 1(1/2^-)$

Mass m = 1730 to 1800 ($\approx$ 1750) MeV

$$\Sigma(1750)\ \text{½}^-$$
$$?$$

```
2       o | ● −
1       o | o
0         |
```

$$\underline{1\text{½}-1\text{½} \mid -\text{½}}$$
$$|$$

$\Sigma(1775)$ 5/2⁻ $\qquad$ $I(J^P) = 1(5/2^-)$

Mass m = 1770 to 1780 ($\approx$ 1775) MeV

$$\Sigma(1775)\ 5/2^-$$
$$?$$

```
2        − | ●
1       −o | o−
0          |
```

$$\underline{1-2 \mid -2\text{½}}$$
$$|$$

$\Sigma(1915)$ 5/2⁺ $\qquad$ $I(J^P) = 1(5/2^+)$

Mass m = 1900 to 1935 ($\approx$ 1915) MeV

$$\Sigma(1915)\ 5/2^+$$
$$?$$

```
2        o | ● +
1       +o | o+
0          |
```

$$\underline{2-1 \mid 2\text{½}}$$
$$|$$

137

$\Sigma(1940)\ 3/2^-$ $I(J^P) = 1(3/2^-)$

Mass m = 1900 to 1950 ($\approx$ 1940) MeV

$\Sigma(1940)\ 3/2^-$

?

```
 2      + | ●
 1     +o | o+
 0        |

      -3 | -1½
         |
```

$\Sigma(2030)\ 7/2^+$ $I(J^P) = 1(7/2^+)$

Mass m = 2025 to 2040 ($\approx$ 2030) MeV

$\Sigma(2030)\ 7/2^+$

?

```
 2      o | ●+
 1     +o | o+
 0        |

   2½-½ | 3½
        |
```

$\Sigma(2250)$ $I(J^P) = 1(?^?)$

Mass m = 2210 to 2280 ($\approx$ 2250) MeV

$\Sigma(2250)$

?

```
 2      o | ●+
 1     +o | o+
 0        |

      3 | 4½     Unknown
        |
```

Ξ⁰ $I(J^P) = \frac{1}{2}(\frac{1}{2}^+)$ P is not yet measured; + is the quark model prediction.
Mass m = 1314.86 ± 0.20 MeV

```
              Ξ⁰
              ?
  2        o | ●+
  1      −+o | +−
  0          |

    1½−1½ | ½
          |
```

Ξ⁻ $I(J^P) = \frac{1}{2}(\frac{1}{2}^+)$ P is not yet measured; + is the quark model prediction.
Mass m = 1321.71 ± 0.07 MeV

```
              Ξ⁻
              ?
  2        −+ | ●+
  1      −+o | +−
  0          |

    1½−1½ | ½
          |
```

Ξ(1530)⁰ 3/2⁺ $I(J^P) = \frac{1}{2}(3/2^+)$

Mass m = 1531.80 ± 0.32 MeV   (S = 1.3)

```
         Ξ(1530)⁰ 3/2⁺
              ?
  2        o | ●+
  1      −+o | +−
  0          |

    2−1 | 1½
        |
```

139

$\Xi(1530)^{-}$ $3/2^{+}$ $I(J^P) = \frac{1}{2}(3/2^{+})$

    Mass m = 1535.0 ± 0.6 MeV

```
               Ξ(1530)⁻ 3/2⁺
                    ?
     2        +○ | ●
     1        −+ | +−
     0            |

              2−1 | 1½
                  |
```

$\Xi(1690)$ $I(J^P) = \frac{1}{2}(?^{?})$

    Mass m = 1690 ± 10 MeV

```
               Ξ(1690)
                    ?
     2        +○ | ●+
     1        −+ | +−
     0            |

              2−1 | 2        Unknown
                  |
```

$\Xi(1820)$ $3/2^{-}$ $I(J^P) = \frac{1}{2}(3/2^{-})$

    Mass m = 1823 ± 5 MeV

```
               Ξ(1820) 3/2⁻
                    ?
     2         ○ | ●−
     1        −+○ | ○+−
     0             |

              1−2 | −1½
                  |
```

140

$\Xi(1950)$ $I(J^P) = \frac{1}{2}(?^?)$

Mass m = 1950 ± 15 MeV

$\Xi(1950)$

?

```
2    +o | ●+
1    −+o | o+−
0        |
```

```
1½−1½ | 1    Unknown
        |
```

$\Xi(2030)$ $I(J^P) = \frac{1}{2}(\geq 5/2^?)$

Mass m = 2025 ± 5 MeV

$\Xi(2030)$

?

```
2    +o | ●+
1    −+o | o+−
0        |
```

```
2−1 | 2    Unknown
      |
```

Ω BARYONS
(S = − 3, I = 0)

Ω^- $I(J^P) = 0(3/2^+)$ J^P is not yet measured; $3/2^+$ is the quark model prediction.

Mass m = 1672.45 ± 0.29 MeV

Ω^-

?

```
2    +o | ●
1    − | +
0       |
```

```
2−1 | 1½
     |
```

141

$\Omega(2250)^-$ $I(J^P) = 0(?^?)$

Mass m = 2252 ± 9 MeV

$\Omega(2250)^-$

?

```
2     o | ●+
1     − | +
0       |
```

1½−1½ | ½   Unknown

|

CHARMED BARYONS
(C = + 1)

Λ_c^+ $I(J^P) = 0(\frac{1}{2}^+)$ J not confirmed; ½ is the quark model prediction.

Mass m = 2286.46 ± 0.14 MeV

Λ_c^+

?

```
3       | o
2       | ●
1     + |
0       |
```

1−1 | ½

|

$\Lambda_c(2595)^+$ $I(J^P) = 0(\frac{1}{2}^-)$ The spin-parity follows from the fact that $\Sigma_c(2455)\pi$ decays, with little available space, are dominant.

Mass m = 2592.25 ± 0.28 MeV

$\Lambda_c(2595)^+$

?

```
3     + | +
2       | ●
1     + |
0       |
```

1−2 | ½

|

142

$\Lambda_c(2625)^+$ $I(J^P) = 0(3/2^-)$ J^P is expected to be $3/2^-$.

    Mass $m = 2628.11 \pm 0.19$ MeV $(S = 1.1)$

$$\Lambda_c(2625)^+$$

```
           ?
3        - | -
2        _ | •
1        - | o
0          |

     1½-1½ | -1½
           |
```

$\Lambda_c(2880)^+$ $I(J^P) = 0(5/2^+)$ There is some good evidence that

indeed $J^P = 5/2^+$.

    Mass $m = 2881.53 \pm 0.35$ MeV

$$\Lambda_c(2880)^+$$

```
           ?
3        + | +
2        _ | •
1        + | o
0          |

     2-1 | 2½
         |
```

$\Lambda_c(2940)^+$ $I(J^P) = 0(?^?)$

    Mass $m = 2939.3^{+1.4}_{-1.5}$ MeV

$$\Lambda_c(2940)^+$$

```
           ?
3        + | +
2        _ | •
1        + | o
0          |

     2-1 | 2½        Unknown
         |
```

143

$\Sigma_c(2455)^{++}$ $I(J^P) = 1(1/2^+)$ J^P not confirmed; $1/2^+$ is the quark model prediction.
Mass m = 2453.98 ± 0.16 MeV

$$\Sigma_c(2455)^{++}$$
?

```
3        | o
2      o | ●+
1        |
0        |
```

```
1-1 | ½
    |
```

$\dot\Sigma_c(2455)^+$ $I(J^P) = 1(1/2^+)$ J^P not confirmed; $1/2^+$ is the quark model prediction.
Mass m = 2452.9 ± 0.4 MeV

$$\Sigma_c(2455)^+$$
?

```
3      + | +
2      o | ●+
1        |
0        |
```

```
1-2 | ½
    |
```

$\Sigma_c(2455)^0$ $I(J^P) = 1(\frac{1}{2}^+)$ J^P not confirmed; $1/2^+$ is the quark model
Mass m = 2453.74 ± 0.16 MeV

$$\Sigma_c(2455)^0$$
?

```
3      o | o
2      + | ●
1        |
0        |
```

```
1½-1½ | ½
      |
```

$\Sigma_c(2520)^{++}$ $I(J^P) = 1(3/2^+)$

Mass m = 2517.9 ± 0.6 MeV   (S = 1.6)

$\Sigma_c(2520)^{++}$

?

3	| ○
2	○ | ● +
1	+ | +
0	|

1½–1½ | 1½

|

$\Sigma_c(2520)^0$ $I(J^P) = 1(3/2^+)$

Mass m = 2518.8 ± 0.6 MeV   (S = 1.5)

$\Sigma c(2520)^0$

?

3	+ | +
2	+ | ●
1	+ | +
0	|

1–2 | 1½

|

$\Sigma_c(2800)$ $I(J^P) = 1(?^?)$ Unknown structures.

$\Sigma_c(2800)^{++}$ mass m = 2801^{+4}_{-6} MeV

$\Sigma_c(2800)^+$ mass m = 2792^{+14}_{-5} MeV

$\Sigma_c(2800)^0$ mass m = 2806^{+5}_{-7} MeV   (S = 1.3)

145

. Ξ_c^+ $I(J^P) = \frac{1}{2}(\frac{1}{2}^+)$ $I(J^P)$ not confirmed; $\frac{1}{2}(\frac{1}{2}^+)$ is the quark

model prediction.

Mass m = $2467.8^{+0.4}_{-0.6}$ MeV

$$\Xi_c^+$$

?

```
3        o | o
2        o | ● +
1       - + | + -
0          |
```

$$\underline{1\frac{1}{2} - 1\frac{1}{2} \mid \frac{1}{2}}$$
$$|$$

. Ξ_c^0 $I(J^P) = \frac{1}{2}(\frac{1}{2}^+)$ $I(J^P)$ not confirmed; $\frac{1}{2}(\frac{1}{2}^+)$ is the quark model

Mass m = $2470.88^{+0.34}_{-0.80}$ MeV

$$\Xi_c^0$$

?

```
3        + | +
2        o | ● +
1       - + | + -
0          |
```

$$\underline{1 - 2 \mid \frac{1}{2}}$$
$$|$$

$\Xi_c^{'+}$ $I(J^P) = \frac{1}{2}(\frac{1}{2}^+)$ $I(J^P)$ not confirmed; $\frac{1}{2}(\frac{1}{2}^+)$ is the quark model

prediction.

Mass m = 2575.6 ± 3.1 MeV

$$\Xi_c^0$$

?

```
3        - | -
2        o | ● +
1       - + | + -
0          |
```

$$\underline{1\frac{1}{2} - 1\frac{1}{2} \mid -\frac{1}{2}}$$
$$|$$

146

$\Xi_c'^0$ $I(J^P) = \frac{1}{2}(\frac{1}{2}^+)$ $I(J^P)$ not confirmed; $\frac{1}{2}(\frac{1}{2}^+)$ is the

quark model prediction.
    Mass m = 2577.9 ± 2.9 MeV

$$\Xi_c^0$$

	?
3	○ | ○
2	+ | ●
1	−+ | +−
0	|

1½−1½ \| ½
|

$\Xi_c(2645)^+$ $I(J^P) = \frac{1}{2}(3/2^+)$
    Mass m = $2645.9^{+0.5}_{=0.6}$ MeV   (S = 1.1)

$$\Xi_c(2645)^+$$

	?
3	○ | ○
2	○ | ●+
1	−+ | +−
0	|

2−1 \| 1½
|

$\Xi_c(2645)^0$ $I(J^P) = 1/2(3/2^+)$

    Mass m = 2645.9 ± 0.5 MeV

$$\Xi_c(2645)^0$$

	?
3	○ | ○
2	+ | ●
1	−+○ | ○+−
0	|

2−1 \| 1½
|

$\Xi_c(2790)^+$ $I(J^P) = 1/2(1/2^-)$

J^P has not been measured; $\frac{1}{2}^-$ is the quark model prediction.

Mass m = 2789.1 ± 3.2 MeV

$\Xi_c(2790)^0$

?

```
3        o | o
2        o | • —
1       — + o | o + —
0          |
```

1½ – 1½ | –½
|

$\Xi_c(2790)^0$ $I(J^P) = 1/2(1/2^-)$

J^P has not been measured; $\frac{1}{2}^-$ is the quark model prediction.

Mass m = 2791.8 ± 3.3 MeV

$\Xi_c(2790)^0$

?

```
3        o | o
2        — | •
1       — + o | o + —
0          |
```

1½ – 1½ | –½
|

$\Xi_c(2815)^+$ $I(J^P) = 1/2(3/2^-)$

J^P has not been measured; $\frac{1}{2}^-$ is the quark model prediction.

Mass m = 2816.6 ± 0.9 MeV

$\Xi_c(2815)^0$

?

```
3        o | o
2        o | • —
1       — + o | o + —
0          |
```

1 – 2 | –1½
|

$\Xi_c(2815)^0$ $I(J^P) = 1/2(3/2^-)$

J^P has not been measured; $\frac{1}{2}^-$ is the quark model prediction.

Mass m = 2819.6 ± 1.2 MeV

```
                Ξc(2815)0
                    ?
    3          o | o
    2          - | ●
    1        -+o | o+-
    0            |

           1-2 | -1½
              |
```

$\Xi_c(2980)$ $I(J^P) = 1/2(?^?)$ Unknown structures.

$\Xi_c(2980)^+$ m = 2971.4 ± 3.3 MeV   (S = 2.1)

$\Xi_c(2980)^0$ m = 2968.0 ± 2.6 MeV

$\Xi_c(3080)$ $I(J^P) = 1/2(?^?)$ Unknown structures.

$\Xi_c(3080)^+$ m = 3077.0 ± 0.4 MeV

$\Xi_c(3080)^0$ m = 3079.9 ± 1.4 MeV   (S = 1.3)

$.\Omega_c^0$ $I(J^P) = 0(\frac{1}{2}^+)$ $I(J^P)$ not confirmed; $0(\frac{1}{2}^+)$ is the quark
model prediction.

Mass m = 2695.2 ± 1.7 MeV   (S = 1.3)

```
                Ω 0
                  c
                  ?
    3          o | o
    2          + | ●
    1        -+o | o+-
    0            |

         1½-1½ | ½
              |
```

149

$.\Omega_c(2770)^0$ $I(J^P) = 0(3/2^+)$ $I(J^P)$ not confirmed; $0(\frac{1}{2}^+)$ is the quark

model prediction.

Mass m = 2765.9 ± 2.0 MeV   (S = 1.2)

$$\Omega^0_c$$

?

```
3        o | o
2       +o | ●o
1       -o | o+
0          |
```

2-1 | 1½

─────────────────────────────────────

BOTTOM BARYONS
(B = - 1)

$.\Lambda^0_b$ $I(J^P) = 0(1/2^+)$ $I(J^P)$ not yet measured; $0(1/2^+)$ is the

quark model prediction.

Mass m = 5619.4 ± 0.6 MeV

$$\Lambda^0_b$$

5425.878 398 MeV

```
4        | o
3        |
2       o | ●
1       + |
0        |
```

1½-1½ | ½

─────────────────────────────────────

$.\Sigma^+_b$ $I(J^P) = 1(1/2^+)$ I, J, P need confirmation.

Mass m = 5811.3 ± 1.9 MeV

$$\Sigma^+_b$$

?

```
4       o | o
3         |
2       o | ●+
1         |
0         |
```

1½-1½ | ½

─────────────────────────────────────

. Σ_b^- $I(J^P) = 1(1/2^+)$ I, J, P need confirmation.

         Mass m = 5815.5 ± 1.8 MeV

$$\Sigma_b^-$$

?

```
4     o | o
3     + |
2     o | ●
1       |
0       |
```

1½–1½ | ½
|

. Σ_b^{*+} $I(J^P) = 1(3/2^+)$ I, J, P need confirmation.

         Mass m = 5832.1 ± 1.9 MeV

$$\Sigma_b^{*+}$$

?

```
4     o |
3       |
2       | ● +
1     + | +
0       |
```

1½–1½ | 1½
|

. Σ_b^{*-} $I(J^P) = 1(3/2^+)$ I, J, P need confirmation.

         Mass m = 5835.1 ± 1.9 MeV

$$\Sigma_b^{*-}$$

?

```
4     o |
3       |
2     + | ●
1     + | +
0       |
```

1½–1½ | 1½
|

151

Ξ_b^0 $I(J^P) = 1/2(1/2^+)$ I, J, P need confirmation.

Mass m = 5788 ± 5 MeV

$$\Xi_b^0$$

?

```
4      o | o
3        |
2      + | ●
1     -+ | +-
0        |
```

1½−1½ | ½
      |

Ξ_b^- $I(J^P) = 1/2(1/2^+)$ I, J, P need confirmation.

Mass m = 5791.1 ± 2.2 MeV

$$\Xi_b^-$$

?

```
4      o | o
3        |
2     +o | ●
1     -+ | +-
0        |
```

1½−1½ | ½
      |

GRAVITONS

$g^\pm$ I(J) = 0(2)

Mass not measured yet.

```
        g+                    g-
  About 70.0 MeV       About 70.0 MeV
  2     |               2     |
  1    + | +            1    - | -
  0     |               0     |

     1 | 2                 -1 | -2
       |                      |
```

Higher energy state orbiting particle pairs can also compose $g^\pm$ gravitons.

According to chonomic decay schemes, a 1 spin graviton also exists.

$$I(J) = 0(1)$$

Mass not measured yet.

g°

```
About 70.0 MeV
2     |
1    o | o
0     |
```

```
1 | 1
  |
```

$$I(J) = 0(2)$$

Mass not measured yet.

g°⁺

```
About 70.0 MeV
2       |
1    +o | o+
0       |
```

```
-3 | -2
   |
```

A negative + echon orbits a positive o echon with -1 orbital spin, forming a neutrino. A positive + echon orbits a negative o echon with -1 orbital spin, forming an anti-neutrino. The neutrino and anti- neutrino orbit each other with an additional -1 orbital spin. The result is a g°⁺ graviton with -2 spin.

$$I(J) = 0(2)$$ The spin conjugate graviton g°⁻ is also possible.

g°⁻

```
About 70.0 MeV
2      |
1    -o | o-
0      |
```

```
3 | 2
  |
```

MAGNETONS

Like gravitons, all magnitons are massive particles—either mesons or baryons. They have an echon pair of either −+ or +−

in the lowest non-definition state of 1 and 1 for the B and N values. In the $--$ or $++$ states in gravitons, the echons have opposite magnetic fields which cancel out. In the magnetons one echon is tipped upside down, making the particle more massive, but also making the magnetic fields of the sub-particles not cancel out, but add to a magnetic dipole, which can align with other ambient magnetic dipoles to make magnetic lines of force. Depending on the density of the magnetic dipoles in the aether, there is a limit to how strong a magnetic force can be achieved. The following is the lowest mass magneton:

ω(782)　　　　$I^G(J^{PC}) = 0^-(1^{--})$

Mass m = 781.94 ± 0.12 MeV　(S = 1.5)

ω(782)

```
        ?
2       |
1     - | +
0       |

      1 | 1
        |
```

More magnetons are ω(1420), ω(1650), ω_3 (1670), K_1(1270), J/ψ (1S), ψ (2S), ψ (3770), χ (3872), Y(10860), N(1535)½⁻, Λ (1810)½⁺, Ω⁻, Ω (2250)⁻, and Ω (2770)⁰. We have the measured masses of all those particles already (see earlier in this chapter). Soon (like maybe in six months) we will have the calculated and predicted masses of each particle from first principles also in *Predicting the Masses*, Volume 2, Predicting the Mesons and *Predicting the Masses*, Volume 3, Predicting the Baryons. They won't be much different than the measured values in this chapter.

Chapter 9

Postulates

A.   Short Cut Postulates. A number of postulates are useful in the overall science of the Unified Field Theory and Unified Particle Theory—including the postulates for calculating the masses of elementary particles.

1.   Parsimony Principle: The Universe is constructed according to the simplest design possible to account for the many varied natural phenomena. [1] (Chapter 6).

2.   The author learned how to unite and predict many of the constants, forces, and particles, given but one simple formula:

$$\frac{0 \; arbitrary \; mass \; unit}{0} \equiv M_0 (arb. \; ma. \; u.), \; [1] \; (Chapter \; 8) \qquad (9\text{-}1)$$

M_0 is the strong mass of a whole particle in the relativistic frame as seen by an observer at rest, 0 in the numerator on the left side of the equation is the mass of a whole particle (uniton) at rest, and 1/0 is the gamma factor transforming the non-relativistic frame to the relativistic frame when the uniton travels at exactly the speed of light.   Many fundamental constants, forces, and particles in the Universe can be derived from this simple definition.

3.   The one measured mass that needs to be input in this model is the mass of the electron: The mass of the electron was previously measured to be 0.510 998 928(11) MeV. [1](p. 434).

4. The spin relation (solved for r) for electrons orbiting atoms is: $n\hbar/m_e v$ [2].   (9-2)

5. The spin relation (solved for r) for the electron family of particles is: $r = n\hbar/b^2 m_e c$. [3]   (9-3)

6. The spin relation (solved for r) for the pion family of particles is: $r = n^2 \hbar / b m_e c$. [3] (9-4)

7. Instead of the spin relation of the electron family member intimately involved with the dot and the neutron overall binding orbit, the spin relation (solved for r) for the neutron family of particles and their captive electron is: $r = N \hbar / B m_e c$. [3]. (9-5)

8. Pauli Exclusion Principle applied to sub-particles in particles: No two sub-particles in a particle (except a briefly temporary quasi-particle) can have the same set of parameters. Also, no two mass factors in a particle can have the same j subscript.

9. Conservation law of particle parameters: Once a set of sub-particle parameters is determined, they must stay the same throughout the sub-particle calculations (except where modified by a chonomic process).

10. The states of the meso-electric terms can be calculated by taking particle number p minus 1 for the pion family, p plus 0 for the positron family, and p plus 1 for the neutron family times the corresponding n times $\pi\alpha$.

11. Masses of particle family members terms can be obtained by evaluating the unitless exponential polynomial (obtained by solving for the orbital v_o^2) times the unitless Energy Factor (EF) times the unitless particle g/2 factor times the mass of the electron m_e. We don't know why, but it seems to work. The m and r in the balancing equations are the same as in postulate A.6

12. Except for particles containing a dot ($\bullet$), particle term masses must be summed with all previous mass terms (previous mass terms must be added only once for each new mass term.).

B. Original Set of Postulates in *Electrino Physics.* Science is based on postulates—unproven assumptions. Yet they can be well chosen based on experimental evidence. Einstein's Special Relativity is based on two postulates: 1) the principle of relativity (from which is gleaned the covariance of physical laws); and 2) the constancy of c (the speed of light) to all observers. The following aether model of quasi-relativity is based on postulates also. Aether quasi-relativity postulates are not Einstein's postulates. (The covariance of physical laws and the constancy of c are derived results in aether quasi-relativity.) The postulates for aether quasi-relativity are more fundamental in nature:

Special Quasi-Relativity in an Aether Postulates

1. There is an absolute space mediated by an electrical aether.

2. For every whole particle relativistic action there is an equal and opposite reaction.

The first postulate may be expanded to say there is a turbulent absolute space mediated by an electrical aether affected by matter and forces in the Universe. The aether may be electrically polarized, ionized, magnetically aligned, and accelerated relative to the distant stars. The best evidence is the aether consists of a sea of imaginary mass integral spin bosons.

The second postulate (parallel to Newton's third law) is here taken more broadly. Lenz's law with magnetic fields and electrical currents ("the induced emf is always in such a direction as to produce a current and resulting flux change that counteracts the original flux change responsible for the induction") might be considered a phenomenon explained by a more broad interpretation of Newton's third law. Nature resists the collapse of a magnetic field by the induction of a current just equal to that which would compensate for the loss of the magnetic field. Aether relativity depends on a similar reaction to an action affecting the permittivity of free space, length, and clock speed of an object due to motion of that object in an aether in a "classical mechanics" system.

General quasi-relativity requires an additional postulate over the two postulates of special quasi-relativity.

General Quasi-Relativity Postulate

3. All non-zero mass particles (including imaginary mass particles) accelerate at the same rate at the same potential in a gravitational field.

In addition to the three postulates of Special and General Quasi-Relativity in an Aether, several postulates are necessary to derive the structure of elementary particles and unite the forces. The postulates will be listed here and discussed where appropriate in the text and derivation.

Particle and Field Postulates

4. Parsimony Principle: The Universe is constructed according to the simplest design possible to account for the many varied natural phenomena.

5. A symmetric smooth charge distribution cannot have detectable spin.

6. (ad hoc hypothesis, not a postulate) In every particle other than zero mass, electrinos have a velocity component along one direction equal to or greater than c. Electrinos may also have velocity components or zero in perpendicular directions.

7. Total momentum P, observable angular momentum $J\hbar$, and total ordinary energy W are conserved in every natural non-accelerating frame and reaction.

8. The observable angular momentum (spin) of the parameterized particle system is

$$s_p = \left[r_p \, x \left(m_{f_p} c_t \right) \right] = \sum_i \left[f_p \hbar / 2 \right]_i 1_J, \qquad (9\text{-}6)$$

in the aether rest frame, where the variables are defined and illustrated in the text.

Definitions of Postulates' Variables and Terms

In hypothesis 6, c is the ordinary speed of light.

In postulate 7, the word "natural" is employed because the postulate holds true for natural particles—those made of unitons, semions, and quartons and their anti-particles in matter, light, and gravitons. Octons, however, are supernatural particles, which, while they may conserve order energy (positive and negative energy in the creation of particles), they do not conserve entropy energy (ordinary energy W—the absolute value, term for term, of order energy in the equations).

In postulate 8, s_p is the observable spin of the overall particle or particles studied. $s_p = J\hbar$. Σ_i means sum over all the sub-particles being considered. r_p means the radius vector of the overall particle contemplated. x means the vector cross product. m_{fp} means the mass of an electrino i making up the overall particle p considered. For an electron it's the mass of a semion. For a photon it's the mass of a uniton. In the relativistic frame, it is the strong mass of the electrino (to be derived). In the nonrelativistic frame, it is a portion of the overall particle considered (such as half of the electron). c_t means the tangential (perpendicular to the radius) or orbital vector velocity at the speed of light. f_p equals the fraction of a whole particle for electrino i in overall particle p, or the fusion state of the electrino. f_p are fractions of charge, and come in +1, -½, +¼, -⅛, 0, +⅛, -¼, +½, and -1 only. The first four are matter, and the last four are antimatter. $\hbar$ equals Planck's constant h divided by 2π. 1_J is a unit vector in the direction of the angular momentum $J\hbar$.

Equation (9-6) holds true in the hyperoptic speed frame as well as the suboptic speed frame. In the hyperoptic speed frame there may be imaginary radii, minus imaginary mass, and positive or negative v_t. The cross product is real and obeys the right hand side

159

of the equation. Particle systems are mass singularities. Because the observer cannot see across to the back side of a mass singularity, but can observe only the front side at the event horizon, s_p is the product of r_p, m_{fp}, and c_t.

Equation (9-6) holds true in the aether rest frame. That is as if an observer rode with the aether particles from at rest at infinity to the surface of the particle and made the observation as he/she passed the surface of the particle on the way in, or rode with the aether particles from at rest at infinity through the particle, and observed at the surface of the particle on the way out. The aether rest frame is not the customary, natural, or convenient frame from which to observe and calculate. It is more natural to select the echon relative rest frame. That frame is at rest with the center of mass of the particle. In that frame the aether travels with velocity v radially. Calculating the angular momentum or spin in this frame requires the addition of terms as shown in Equation (9-7).

$$s_p = \left[r_p \, x \left(m_{f_p} V_p \right) \right] = \sum_i \left[f_p \frac{1}{2} \hbar k_p \right]_i 1_J \qquad (9\text{-}7)$$

V_p is different than c_t in Equation (9-6). For the electron, it is the vectorial sum of c_t and v_p, which are the orbital velocity of the electrino and the radial velocity of the aether at the surface of particle p, respectively. k_p is added. $k_p = (1 + v_p^2/c^2)^{1/2}$ and is on the order of $1 + 10^{-45}$ for electrons and differs negligibly from 1 for conceivable v_p. v_p is v for particle p, the magnitude of v_p. Equation (9-6) is a good approximation for equation (9-7).

The Parsimony Principle is a long-standing physical view of the Universe. It has been employed in many theories. This principle is the foundation for this derivation of particle and field structure. Some of the conservation laws in Postulate 7 have long been known and used in physics. The rest of the postulates are unfamiliar ones introduced by the author.

160

[1] Gordon L. Ziegler, *Electrino Physics* (http://benevolententerprises.org Book List; 11/23/2013, Xlibris LLC). (*Electrino Physics* Draft 2 now available at Amazon.com, CreateSpace.com and Kindle.)

[2] Gordon L. Ziegler and Iris Irene Koch, "Prediction of the Masses of Charged Leptons," Galilean Electrodynamics **20** (6), 114-118 (2009).

[3] Gordon L. Ziegler and Iris Irene Koch, "Prediction of the Masses of Every Particle, Step 1," **Galilean Electrodynamics**, Summer 2010, Vol. 21, SI No.3, pp. 43-49.

Chapter 10

The Science of Calculating the Masses of Elementary Particles

It is appropriate now to explain the science of calculating the masses of elementary particles from the information in their chonomic structures. This work was done for charged leptons, the pion family, and the neutron family in [1,2]. (The neutron family and pion family calculations are modified at the end of this paper.) The first family was of particles composed simply of orbiting fractons in containment. The pion and neutron family calculations are more complex, but are good explanations of the science of calculating the masses of systems in the chonomic structures.

In order to abbreviate the requisite labor in the calculations involved in this book and take advantage of previous results, we will reverse some of the order of the calculations. Instead of combining the v_o^2 and solving for m, we will henceforth solve for the m's and combine masses employed in the particles. Some of the masses will be masses previously calculated in "Prediction of the Masses of Every Particle, Step 1." [2] These will have to be updated slightly because of the update of the α to 2010 CODATA data. Some of the masses will be masses associated with the potential energies and kinetic energies of the orbits binding the masses together. The previously calculated masses include their g/2 factors and energy factors combining their kinetic and their potential energies. We will have to include both terms in the calculations of each intrinsic mass and orbital mass associated with the particles on the face of the chonomic grid for the particle in question.

The particle data below are formatted similarly for each particle: first on the left is a particle symbol. To the right is typically the isospin, spin, parity, etc. Below that is the measured mass. Then the particle symbol is repeated above the particle chonomic structure proper, as part of the particle structure. Then comes another mass. Originally it was the measured mass less the error terms. But we desire to make a change here—let us now make this mass the calculated mass from first principles. Below that is the chonomic grid. Right of the vertical line are the symbols of positively charged particles—to the left the negative particles. There are different levels in the grid representing different energy states. For more information, see [3] Chapter 10.

η

$$I^G(J^{PC}) = 0^+(0^{-+})$$

Mass m = 547.853 ± 0.024 MeV

η

548.008 806

```
3    |
2    |
1  -+|+-
0    |

1½-1½|0
     |
```

This is the chonomic structure for the η particle, our first sample particle for mass calculation. First let us calculate the masses of the intrinsic particles on the face of the grid. (Then we will do the binding orbital masses.) They are all muons and anti-muons, which are in the electron family and the positron family, which we have to calculate in order to obtain the muon and anti-muon mass values.

A. Electron Family Calculations

Deriving particle states is one orbital level deeper than deriving electron orbital states that Niels Bohr did. The calculations are similar, but significantly different. Instead of treating the situation as a one body problem, we must treat the situation as a two body problem, with two equal semions in orbit about each other. [4] This introduces an extra ½ into the expression of the centrifugal force.

We will solve for the particle states in a different order than Neils Bohr did. First we will balance the strong electric force in this problem with the central force of inertia in this problem, and familiarize ourselves with the whole problem to be solved:

Equation (10-1) below is the balancing of the force due to charge on the semions with the centrifugal force on the semions. The effective mass of a semion is half the mass of the whole particle in the outer non-relativistic frame. We use this mass of the semion in the centrifugal force along with the ½ from the two body problem. The velocity v_o is greater than or equal to c, and must increase when the energy increases. In the electric type force side of the equation, the charge of the semion is e/2.

Different particle systems are in different order black holes. The force must depend on the order of black hole the particle system is in. Like the strong force and the electric force differ in strength by a power of $1/\alpha$, the forces in different orders of black holes differ by powers of $1/\alpha^{n/b}$. The electric type force expression, in the right side of Eqn. (10-1), we expect to depend on a power of $1/\alpha^{n/b}$. To be in harmony with measured results and Eqn. (10-1), we want the power of $1/\alpha$ to be related to n/b. Also, we want the power for the electron to be such that the power of α is 1 when n = 0. We therefore take the power of α for electrons and higher charged leptons to be n/b + 1. Completing the balancing of forces equation, we have

$$\frac{1}{2}\frac{m_e}{2}\frac{v_o^2}{r} = \left(\frac{e}{2}\right)^2 \frac{1}{4\pi\varepsilon_0 \alpha^{(n/b)+1}r^2} \tag{10-1}$$

The first ½ in the equation is from the two body nature of the problem, converting it to a one body problem. The $m_e/2$ is for the semion in ground state. The v_o^2/r is the centrifugal force acting on the semions from a circular orbit. The e/2 is for the charge on each semion. The $4\pi\varepsilon_0$ are constants necessary to solve this problem in MKSC units. The $1/\alpha^{n/b}$ is the fine structure coupling constant between the orders of black holes in the problem. The r is the radius of the particle sub-particle orbit.

Thanks to the two body nature of the problem, all of the numeral constants except 4 in the above equation cancel out. And $e^2/4\pi\varepsilon_0\alpha$ can be factored out as ℏc. One r can cancel out of the two sides of the equation. The equation then looks like the following:

$$m_e v_o^2 = \frac{\hbar c}{\alpha^{n/b} r} \qquad (10\text{-}2)$$

We can solve for v_o if we can find an independent equation for r. Neils Bohr utilized the spin relation for electrons for this purpose. That equation was

$$m_e vr = n\hbar. \qquad (10\text{-}3)$$

Unfortunately that spin relation does not work for charged leptons. Through trial and error, the author has settled on the following spin relation for charged leptons, which is here taken as a postulate:

$$m_e cr = n\hbar/b^2. \qquad (10\text{-}4)$$

Solving for r we obtain:

$$r = n\hbar/b^2 m_e c. \qquad (10\text{-}5)$$

Combining Eqn. (10-2) with Eqn. (10-5) we have:

$$m_e v_o^2 = \hbar c b^2 m_e c / \alpha^{n/b} n\hbar \qquad (10\text{-}6)$$

$$v_o^2 = (b^2/n\alpha^{n/b}) c^2 \qquad (10\text{-}7)$$

$$v_o = (b^2/n\alpha^{n/b})^{1/2} c. \qquad (10\text{-}8)$$

The constant n in the above Eqns. (10-1) – (10-8) is confusing. The n in Niels Bohr solution went 0, 1, 2, 3, 4, 5 . . . , but the n in the charged lepton solution goes 0, 1, 3, 6, 10, 15, 21 . . . , where b in the charged lepton solution goes 0, 1, 2, 3, 4, 5 . . . . The n's are calculable from the b's. $n_j = n_{j-1} + b_j$.

B. Deriving Semion Orbit Energy Levels and Masses

1. Energy calculations.

We have solved for v_0 in terms of our parameters n and b. We can now plug that formula into the relationship for particle energy to obtain the energy levels of quarton orbits, and thus the particle masses. The kinetic, potential, and total energies of the semion system can be expressed as

$$Energy_{total} = Energy_{kinetic} + Energy_{potential}. \tag{10-9}$$

Because, in some particles, like charges attract, the potential energy for those particles is positive instead of negative. You recall the energies are negative in Neils Bohr's calculations of orbiting electrons in Hydrogen. But they are positive in our calculations. Also the kinetic Energy in some particles is a 0 or net 0, but is one half the potential energy in other cases, so that the EF for those cases is 3/2. We must take that into consideration in a term EF (energy factor). But that is not all! The value we would thus calculate is most of a mass term, not the total mass of the particle we are calculating. The mass term must be summed up with all previous mass terms in the series (previous terms utilized only once in each term) to arrive at the total mass of the particle. Also we must multiply by the appropriate g/2 factor in each term to have the complete mass term for summing.

$$m = \left\{ \left(\frac{b_j^2}{n_j \alpha^{n_j/b_j}} \right) EF_j (g/2)_j + \left(\frac{b_{j-1}^2}{n_{j-1} \alpha^{n_{j-1}/bj-1}} \right) EF_{j-1} (g/2)_{j-1} + ...1 \right\} m_e$$

(10-10)

For the calculation of the intrinsic masses of the η particle we must have the muon term of the electron family calculation in Eqn. (10-10) above (only one large term and 1), 3/2 for the EF_j, and the appropriate g/2 factor for each term. Since we are doing all the

166

necessary calculations from first principles, we will take time out here to derive the necessary g/2 factors from first principles:

2. g/2-factor Theory

Our model of half the g-factor is a sum of terms—each term for a force or interaction—each term an integration of a force with respect to r or R (in other words, an energy), divided by the standard energy of that particle system in the strong relativistic frame or the non relativistic frame, as appropriate (in other words, $E = Mc^2$ or $E = mc^2$, respectively).

The forces in the strong relativistic frames we integrate to the standard radius in the inner relativistic frames, namely R. The forces in the non-relativistic frames–electricity and magnetism--we integrate to the standard in non-relativistic frames, namely $2\pi r$. For the weak forces, we integrate to the standard $2\pi R$, because there are no magnetic monopoles. In the following general term derivations, the integration limit will be expressed as lim, to be substituted from Table 9-1. The term lim will either be 1 for the strong force and gravity, or 2π for all other forces. For the magnetic force and weak forces, the $2\pi r$ or $2\pi R$ is for only one r or R in r^2 or R^2 (in other words $r(2\pi r)$, etc.

The inertial force is not included in this calculation. Strong gravity and the strong electric force are equated in the model. They are not additive. They are equivalent calculation of forces. Therefore we need to consider only one of the two—strong electric or strong gravity forces—in half the g-factor. The following is the pattern of terms we will encounter in the calculated g/2 factors.

$$g_p / 2 \equiv \left(\Sigma t_{se} + t_{me} + t_e + t_{mg} + t_{wk1} + t_{wk2} + ... + t_g \right)_p, \quad (10\text{-}11)$$

where there is defined to be the term, se stands for strong electric, me stands for meso-electric, e electric, mg magnetic, wk1 weak$_1$, etc., and g gravity. We will calculate each term separately.

In all the magnetic (and weak) type forces and interactions, the integration is done already. We only have to adjust the resultant energy for the R or r limits and weak terms, calculate, and simplify.

The magnetic and weak g/2 terms have in the calculations a velocity v_2, which in these calculations is orbital. For all subscripted weak interactions, $v_2 = c$. For the magnetic force, $(v_2 \ll c) \approx 0$.

The exponent of α k is the sum of two parameters p and q: p is the sequential number of a list of forces; q is the number of shell the mass singularity is in. As with electron shells, singularities in particles do not all come in a single shell. For electrons and muons, q = 0. For tauons and a few higher charged leptons, q = 1. Be aware that q will switch to 2 and higher numbers as p increases beyond the first few numbers.

Like $v^2 = 2GM/r$ for speeds much, much less than c, and GM/r for speeds about c, the denominators of some g/2 factor magnetic and weak terms have an extra factor of two in them. In the calculations, a parameter Δ substituted from Tables 10-1 or 10-3 accounts for these differences. For the magnetic term and weak$_1$ term for the electron g/2-factor, $\Delta = 2$. For all other weak terms for any particles, $\Delta = 1$. The appropriate values of Δ will be recorded in Tables 10-1 and 10-3.

Please note that r^2 is positive, R^2 is negative, m^2 is positive, M^2 is negative, rm is positive, and RM is positive. In the expression two charges are multiplied together—ee. The test charge is negative—the electron—e. If the main charge is also –e, the negative signs cancel, and the Sign is +. If the main charge is positive, the Sign is -. In the electric force, r and m are used (the outer non-relativistic values). In the strong electric force, R and M are used (the inner relativistic values). In the equations, the radius is abbreviated as rad.

3. Calculation of g/2-factor terms

First we calculate the general terms, employing the above data for the g/2-factor. Then we will calculate the individual force terms for the electron g/2-factor and for the muon g/2-factor.

a. General electric term.

$$t_{ge} \equiv \frac{- \int\limits_{\infty}^{\lim rad} \dfrac{sign\ e^2}{4\pi\varepsilon_0 \alpha^k rad^2}}{mass\ c^2}, \qquad (10\text{-}12)$$

$$= \frac{- \hbar c}{\lim rad\ \alpha^{(k-1)}\ mass\ c^2}, \qquad (10\text{-}13)$$

$$= \frac{- \hbar c\ c}{\lim c^2 \alpha^{(k-1)} \hbar}, \qquad (10\text{-}14)$$

$$= \frac{- 1}{\lim \alpha^{(k-1)}}. \qquad (10\text{-}15)$$

For the electric force, $\lim = 2\pi$, $k = 0$. Therefore, for the electric force, $t_e = -\alpha/(2\pi)$.

For the strong electric force (equivalent to the strong gravitational force), $\lim = 1$, $k = 1$. Therefore $t_{se} = -1$. This is the same as the strong gravitational term t_{sg} calculated below:

b. Strong gravitational term.

$$t_{sg} \equiv \frac{- \int\limits_{r=\infty}^{R_0} \dfrac{GM_0^2}{r^2}\,dr}{M_0 c^2}, \qquad (10\text{-}16)$$

$$= \frac{GM_0^2}{R_0 M_0 c^2} = \frac{- M_0^2}{M_0^2} = -1. \qquad (10\text{-}17)$$

Only one strong term is employed in the g/2-factor.

The explanation of the minus signs in Eqns. (10-16) and (10-17) is as follows: The minus sign is because of the backward integration.

Another minus sign comes from the integration of r^{-2} to r^{-1}. And another minus sign comes from the fact that the strong nuclear force is attractive. The net effect is one minus sign.

c. Magnetic term.

The magnetic force makes a split of the energy state of $e\hbar B/m$ separation when in a magnetic field for electrons in atoms, adding $e\hbar B/2m$ to the maximum energy state in atom orbits. We are interested in how much energy a magnetic field can add to the maximum energy state of a particle. The charge of electrinos is a fraction of e, but the charge sums to e in common particles. Also the relativity reduced mass of the electrinos is a fraction of the mass of the particle m, but the mass sums to m in the particle. Taking the energy $e\hbar B/2m$ without integration, but putting in the limits of the 'would have been' integration, and remembering the Δ term for magnetic type terms, we obtain the magnetic term. In the general electric term, α is always positive. But in the magnetic and weak terms, α is alternately negative or positive, depending on the power p. For the magnetic term, we will take $p \approx 0$ and $k_p \approx 0$. So we also make note of that in the following terms. (The parameters are defined in [1] Chapter 7, pp. 217-221.)

$$t_{mg} = \frac{e\hbar(-1)^P B}{2\Delta \; mass \; mass \; c^2},$$
(10-18)

$$= \frac{e\hbar(-1)^P \alpha^{k_p}}{2\Delta \; mass^2 \; c^2} \frac{\mu_0 e(-v_2)}{4\pi \; rad \; \lim \; rad},$$
(10-19)

$$= \frac{e^2}{4\pi\varepsilon_0 \alpha} \frac{\hbar(-1)^P \alpha^{(k_p+1)}(-v_2)}{2\Delta c^4 \; \lim} \frac{1}{r^2 m^2}.$$
(10-20)

But
$$r^2 m^2 = \frac{\hbar^2}{c^2}.$$
(10-21)

Therefore $t_{mg} = \dfrac{\hbar^2 c^2 \left(-v_2/c\right)\left(-1\right)^p \alpha^{\left(k_p+1\right)} c^2}{\lim 2\Delta c^4 \hbar^2}$, (10-22)

$$= \dfrac{\left(-v_2/c\right)\left(-1\right)^p \alpha^{\left(k_p+1\right)}}{\lim 2\Delta}.$$ (10-23)

But $p = k_p = 0$, $\lim = 2\pi$, *and* $\Delta = 2$, (10-24)

so $\qquad\qquad t_{mg} = \dfrac{\left(-v_2/c\right)\alpha}{8\pi}.$ (10-25)

Where v_2 is $\ll$ c (as in magnetism), it is from one order less mass singularity hole than c is. We already have an example of what one order less singularity does to the g/2 factor term. The electric is one order less a singularity than the strong. The electric term is $-\alpha/(2\pi)$. The strong term is -1 The ratio is $\alpha/(2\pi)$. For the magnetic term, we take $v_2/c = \alpha(2\pi)$. Then $t_{mg} = -\alpha^2/(16\pi^2)$.

d. General weak term.

Weak forces are similar to magnetic forces, except they have more terms to account for magneton mediation and mass distinguishing constant engrossment. Also $v_2/c = 1$ in all cases, so there is an extra (-1) for $(-v_2/c)$ in the terms. The equation for the general weak term is

$$t_{wk} = \dfrac{e\hbar(-1)^p B_{wk}}{2\Delta \; mass \; mass \; c^2},$$ (10-26)

$$= \dfrac{e\hbar(-1)^p \alpha^{k_p}}{2\Delta mass^2 c^2} \; \dfrac{\mu_0 e(-c)n(32)^{k_p-2}}{4\pi \, rad \lim rad},$$ (10-27)

$$= \frac{e^2}{4\pi\varepsilon_0\alpha} \frac{\hbar(-1)^P \alpha^{(k_p+1)}(-c)}{2\Delta c^4 \lim} \frac{n(32)^{k_p-2}}{R^2 M^2}.$$ (10-28)

But $e^2 / 4\pi\varepsilon_0\alpha = \hbar c$, $RM = \hbar / c$, lim $= 2\pi$, and $\Delta = 1$ (10-29)

for weak$_2$ and higher weak forces. Thus the master weak term for weak$_2$ and higher weak forces, where $k_p \geq 2$, is

$$t_{wk} = \frac{1}{4\pi}(-1)(-1^P)n(32)^{k_p-2}\alpha^{k_p+1}.$$

(10-30)

For lower order weak terms it is best to determine the terms through experimental calculations and best fit to the g/2 factor.

e.   Gravity term.

$$t_g \equiv \frac{-\int_\infty^r -\frac{Gm^2}{r^2}dr}{mc^2} = -\frac{Gm^2}{rmc^2},$$ (10-31)

$$= -\frac{G}{\hbar c}m^2 = \frac{m^2}{M_0^2}.$$ (10-32)

The author's guess is that the gravitational term of the g/2-factor also has r limits of integration. It will require 45 place accuracy to disprove this. For exact g/2-factors, the gravity term is required. However, if less than 45 place accuracy is required, the gravity term can be neglected.

f. Meso-electric term.

For positive particle g/2 factors, there is an extra g/2 factor term for the meso-electric force = $-bn\pi\alpha$. This starts at 0 state with the positron with zero value. Thus the positron is the only positive particle that is the simple charge conjugate of its negative particle (the electron). For states 1 or higher, the meso-electric term makes the greatest difference of the g/2 factor terms.

4. Evaluating the Electron g/2-factor

We begin to evaluate the electron g/2-factor by completing Table 10-2.

Table 10-1

Parameters Used in Calculating the Electron g/2-factor Terms

force factor term	sign	rad	mass	Δ	lim	v_2	p	q	k_p	n	electron g/2-factor term
strong	+	R	M	1					1		-1
electric	+	r	m		2π				0		$-\alpha/(2\pi)$
magnetic		r	m	2	2π	≈ 0	0	≈ 0	≈ 0	1	$-\alpha^2/(16\pi^2)$
weak$_1$		R	M	2	2π	c	1	≈ 0	1	1	$\alpha^2/(8\pi)$
weak$_2$		R	M	1	2π	c	2	≈ 0	2	1	$-\alpha^3/(4\pi)$
weak$_3$		R	M	1	2π	c	3	≈ 0	4	1	$(32\alpha)^1\alpha^3/(4\pi)$
weak$_4$		R	M	1	2π	c	4	1	5		$-(32\alpha)^3\alpha^3/(4\pi)$
weak$_5$		R	M	1	2π	c	5	1	6	1	$(32\alpha)^4\alpha^3/(4\pi)$
weak$_6$		R	M	1	2π	c	6	1	7	1	$-(32\alpha)^5\alpha^3/(4\pi)$
weak$_7$		R	M	1	2π	c	7	1	8	1	$(32\alpha)^6\alpha^3/(4\pi)$
weak$_8$		R	M	1	2π	c	8	1	9	1	$-(32\alpha)^7\alpha^3/(4\pi)$
etc.											
gravity		r	m	1							$m_e^2/(M_0)^2$

The right hand column of Table 10-1 can be evaluated if we obtain a value for α, the Fine Structure Constant. The previous (2006) CODATA value was 0.007 297 352 5376(50). The most recent (2010) CODATA value [8] is 0.007 297 352 5698(24). The author used both the CODATA values and the natural units system to evaluate the g/2-factors. For the evaluation of the electron g/2-factor, see Table 10-2 below.

Table 10-2

Electron g/2-factor Evaluation

force	electron g/2-factor term	numerical value (2006)	numerical value (2010)	natural units
strong	-1	$-1.000\ 000\ 000\ 000\ \ldots$	$-1.000\ 000\ 000\ 000\ \ldots$	-1
electric	$-\alpha/(2\pi)$	$-0.001\ 161\ 409\ 727\ \ldots$	$-0.001\ 161\ 409\ 733\ \ldots$	$-1/8\pi^2$
magnetic	$-\alpha^2/(16\pi^2)$	$-0.000\ 000\ 337\ 218\ \ldots$	$-0.000\ 000\ 337\ 218\ \ldots$	$-1/256\pi^4$
weak$_1$	$\alpha^2/(8\pi)$	$+0.000\ 002\ 118\ 804\ \ldots$	$+0.000\ 002\ 118\ 804\ \ldots$	$1/128\pi^3$
weak$_2$	$-\alpha^3/(4\pi)$	$-0.000\ 000\ 030\ 923\ \ldots$	$-0.000\ 000\ 030\ 923\ \ldots$	$-1/128/\pi^4$
weak$_3$	$(32\alpha)^1\alpha^3/(4\pi)$	$+0.000\ 000\ 007\ 221\ \ldots$	$+0.000\ 000\ 007\ 221\ \ldots$	$1/8\pi^5$
weak$_4$	$-(32\alpha)^3\alpha^3/(4\pi)$	$-0.000\ 000\ 000\ 393\ \ldots$	$-0.000\ 000\ 000\ 393\ \ldots$	$2/\pi^7$
weak$_5$	$(32\alpha)^4\alpha^3/(4\pi)$	$+0.000\ 000\ 000\ 091\ \ldots$	$+\ 0.000\ 000\ 000\ 091\ \ldots$	$16/\pi^8$
weak$_6$	$-(32\alpha)^5\alpha^3/(4\pi)$	$-0.000\ 000\ 000\ 021\ \ldots$	$-\ 0.000\ 000\ 000\ 021\ \ldots$	$128/\pi^9$
weak$_7$	$(32\alpha)^6\alpha^3/(4\pi)$	$+0.000\ 000\ 000\ 005\ \ldots$	$+0.000\ 000\ 000\ 005\ \ldots$	$1024/\pi^{10}$
weak$_8$	$-(32\alpha)^7\alpha^3/(4\pi)$	$-0.000\ 000\ 000\ 001\ \ldots$	$-0.000\ 000\ 000\ 001\ \ldots$	$8192/\pi^{11}$
etc.				
gravity	$m_e^2/(M_0)^2$	$-0.000\ 000\ 000\ 000\ \ldots$	$-0.000\ 000\ 000\ 000\ \ldots$	
Total with eight weak interactions		$-1.001\ 159\ 652\ 163\ \ldots$	$-1.001\ 159\ 652\ 169\ \ldots$	

The 2010 calculated theoretical electron g/2-factor is -1.001 159 652 169 . . . . The measured 2010 electron g/2 factor is -1.001 159 652 180 76(27). [8] The 2010 theoretical value differs from the measured value by 1.1×10^{-11}. The 2006 calculated theoretical electron g/2 factor was -1.001 159 652 163 . . . . The measured 2002 CODATA electron g/2-factor was -1.001 159 652 1811(08). The theoretical value differed from the measured value by 1.81×10^{-11}. (The difference with 1998 CODATA data was 2.38×10^{-11}, and with 1986 CODATA data was 5.70×10^{-11}.) The fit of the theoretical values with the measured values continues to get better. Still, the measured value of the electron g/2-factor is more accurate than the measured α used to try to calculate it. We need a more accurate α. In natural units $\alpha = 1/4\pi$ exactly.

5.   Evaluating the Muon g/2-factor

The muon g/2-factor differs from the electron g/2-factor in one way. All but one weak interaction terms are multiplied by n ≈-3 for muons. We evaluate the muon g/2-factor by completing Tables 4 and 5.

Table 10-3

Parameters Used in Calculating the Muon g/2-factor Terms

force	sign	rad	mass	Δ	lim	v_2	p	q	k	n	muon g/2-factor term
strong	+	R	M		1				1		-1
electric	+	r	m		2π				0		$-\alpha/(2\pi)$
magnetic		r	m	2	2π	≈0	0	≈0	≈0	-1	$-\alpha^2/(16\pi^2)$
weak$_1$		R	M	1	2π	c	1	≈0	1	-1	$-\alpha^2/(4\pi)$
weak$_2$		R	M	1	2π	c	2	≈0	2	-3	$+3\alpha^3/(4\pi)$
weak$_3$		R	M	1	2π	c	3	1	4	-3	$-(32\alpha)^1\alpha^3/(4\pi)$
weak$_4$		R	M	1	2π	c	4	1	5	-3	$+3(32\alpha)^3\alpha^3/(4\pi)$
weak$_5$		R	M	1	2π	c	5	1	6	-3	$-3(32\alpha)^4\alpha^3/(4\pi)$
weak$_6$		R	M	1	2π	c	6	1	7	-3	$+3(32\alpha)^5\alpha^3/(4\pi)$
weak$_7$		R	M	1	2π	c	7	1	8	-3	$-3(32\alpha)^6\alpha^3/(4\pi)$
weak$_8$		R	M	1	2π	c	8	1	9	-3	$+3(32\alpha)^7\alpha^3/(4\pi)$
etc.											
gravity		r	m		1						$m_\mu^2/(M_0)^2$

Table 10-4

Muon g/2-factor Evaluation

force	muon g/2-factor term	numerical value (2006)
strong	-1	$-1.000\ 000\ 000\ 000\ \ldots$
electric	$-\alpha/(2\pi)$	$-0.001\ 161\ 409\ 727\ \ldots$
magnetic	$-\alpha^2/(16\pi^2)$	$-0.000\ 000\ 337\ 218\ \ldots$
weak$_1$	$-\alpha^2/(4\pi)$	$-0.000\ 004\ 237\ 608\ \ldots$
weak$_2$	$+3\alpha^3/(4\pi)$	$+0.000\ 000\ 092\ 769\ \ldots$
weak$_3$	$-3(32\alpha)^1\alpha^3/(4\pi)$	$-0.000\ 000\ 021\ 663\ \ldots$
weak$_4$	$+3(32\alpha)^3\alpha^3/(4\pi)$	$+0.000\ 000\ 001\ 181\ \ldots$
weak$_5$	$-3(32\alpha)^4\alpha^3/(4\pi)$	$-0.000\ 000\ 000\ 275\ \ldots$
weak$_6$	$+3(32\alpha)^5\alpha^3/(4\pi)$	$+0.000\ 000\ 000\ 064\ \ldots$
weak$_7$	$-3(32\alpha)^6\alpha^3/(4\pi)$	$-0.000\ 000\ 000\ 015\ \ldots$
weak$_8$	$+3(32\alpha)^7\alpha^3/(4\pi)$	$+0.000\ 000\ 000\ 003\ \ldots$
etc.		
gravity	m_μ^2/M_0^2	$-0.000\ 000\ 000\ 000\ \ldots$

Total with eight weak forces $\qquad$ $-1.001\ 165\ 912\ 489\ \ldots$

The 2010 calculated theoretical numerical values will be the same as the 2006 numerical values, except the electric term will be -0.001 161 409 733 . . . and the total will be -1.001 165 912 495 . . . . The 2010 measured CODATA value (after dividing by 2) is -1.001 165 9209(07). [8] The theoretical value differs from the measured value by 8.4×10^{-09}. The 2006 calculated theoretical muon g/2-factor was -1.001 165 912 489 . . . The 2006 CODATA measured muon g/2-factor is -1.001 165 920 7(06). The theoretical value differs from the measured value by 8.21×10^{-9}.

The electron and muon theoretical g/2-factors are close but not exact fits with the measured g/2-factors. When α and the g/2 factors are known more precisely, it will be interesting to see if the measured values are closer to the theoretical values.

The number of identifiable precise terms in the g/2-factors is greatly increased with this scheme. We have good confirmation of the fundamental forces of the Universe by the g/2-factors. These very accurate calculations are without recourse to renormalization theory of Quantum Electrodynamics. They are without the quark hypothesis. These accurate calculations are one test for a new boson-aether model of physics in which charges are divided into e, e/2, e/4, e/8, 0, -e/8, -e/4, -e/2, and –e. Acknowledgment. I am very grateful to Dr. James G. Gilson for discussions on the electron and muon g/2 factors.

6. Evaluating the Positron g/2 factor

Table 10-5

Positron family

Anti-electron -e_0 g/2 Factor Evaluation with 2010 α

force:	g/2 factor term:	numerical value:
strong	$+1$	+1.000 000 000 000
meso-electric	$-bn\pi\alpha$	-0.000 000 000 000
electric	$+\alpha/2\pi$	+0.001 161 409 733
magnetic	$+\alpha^2/16\pi^2$	+0.000 000 337 218
$weak_1$	$-\alpha^2/8\pi$	-0.000 002 118 804
$weak_2$	$+\alpha^3/4\pi$	+0.000 000 030 923
$weak_3$	$-(32\alpha)^1\,\alpha^3/4\pi$	-0.000 000 007 221
$weak_4$	$+(32\alpha)^3\,\alpha^3/4\pi$	+0.000 000 000 393
$weak_5$	$-(32\alpha)^4\,\alpha^3/4\pi$	-0.000 000 000 091
$weak_6$	$+(32\alpha)^5\,\alpha^3/4\pi$	+0.000 000 000 021
$weak_7$	$-(32\alpha)^6\,\alpha^3/4\pi$	-0.000 000 000 005
$weak_8$	$+(32\alpha)^7\,\alpha^3/4\pi$	+0.000 000 000 001
$weak_9$	$-(32\alpha)^8\,\alpha^3/4\pi$	-0.000 000 000 000

total calculated g/2 factor for $-e_0$ +1.001 159 652 169

7. Evaluating the Anti-Muon g/2 Factor

Table 10-6

Positron family

Anti-muon $-e_1$ g/2 Factor Evaluation with 2010 α

force:	g/2 factor term:	numerical value:
strong	$+1$	$+1.000\ 000\ 000\ 00$
meso-electric	$-bn\pi\alpha$	$-0.022\ 925\ 309\ 22$
electric	$+\alpha/2\pi$	$+0.001\ 161\ 409\ 73$
magnetic	$+\alpha^2/16\pi^2$	$+0.000\ 000\ 337\ 21$
$weak_1$	$+\alpha^2/4\pi$	$+0.000\ 004\ 237\ 60$
$weak_2$	$-3\alpha^3/4\pi$	$-0.000\ 000\ 092\ 76$
$weak_3$	$+3(32\alpha)^1\,\alpha^3/4\pi$	$+0.000\ 000\ 021\ 66$
$weak_4$	$-3(32\alpha)^3\,\alpha^3/4\pi$	$-0.000\ 000\ 001\ 18$
$weak_5$	$+3(32\alpha)^4\,\alpha^3/4\pi$	$+0.000\ 000\ 000\ 27$
$weak_6$	$-3(32\alpha)^5\,\alpha^3/4\pi$	$-0.000\ 000\ 000\ 06$
$weak_7$	$+3(32\alpha)^6\,\alpha^3/4\pi$	$+0.000\ 000\ 000\ 01$
$weak_8$	$-3(32\alpha)^7\,\alpha^3/4\pi$	$-0.000\ 000\ 000\ 00$
$weak_9$	$+3(32\alpha)^8\,\alpha^3/4\pi$	$+0.000\ 000\ 000\ 00$
$weak_{10}$	$-3(32\alpha)^9\,\alpha^3/4\pi$	$-0.000\ 000\ 000\ 00$
total calculated g/2 factor for $-e_1$		$+0.978\ 240\ 603\ 27$

8. The Masses of the Muon and Anti-Muon.

We now have all the necessary data to assemble the masses of the muon and anti-muon in Eqn. (10-10) and (10-33):

$$m = \left\{ \left(\frac{b_j^2}{n_j\,\alpha^{n_j/b_j}} \right) EF_j\,(g/2)_j + \left(\frac{b_{j-1}^2}{n_{j-1}\,\alpha^{n_{j-1}/bj-1}} \right) EF_{j-1}\,(g/2)_{j-1} + ...1 \right\} m_e$$

(10-33)

The right most term in Eqn. (10-33) is defined as 1.0.

To calculate Eqn. (10-33) in general, we must have a definition of n, b, and j: (See Table 10-1.) The first three n and b are tested. Higher n and b are calculated. We expect both n and b to increase with j. We expect $n_j - (n_{j-1})$ to be b_j.

Finally, just as the mass is a series of terms, all other force terms are added by multiplying by half the g-factor for the given particle. There are only two usable measured g/2 factors that are available which can be used in calculating masses—for the electron and the muon. Fortunately, the match is close enough between particle masses and particle g/2 factors that calculated g/2 factors can be tested by the measured masses of the particles. We will begin employing calculated g/2 factors in this and the next section.

Particle	b	n	polynomial	EF	g/2 factor	mass factor m/m_e
Electron	0	0	included	included	included	1.000 000 000
Muon	1	1	$1/\alpha$	3/2	-1.001 165 912	205.793 656
					Total Muon	206.793 656
Positron	0	0	included	included	included	-1.000 000 000
Anti-Muon	1	1	$1/\alpha$	3/2	+0.978 165 912	-201.065 914
					Total Anti-Muon	-202.065 914

Table 10-7

9. Masses of + spin muons and anti-muons

The calculations above are for the − spin muons and anti-muons. The g/2 factors of the + spin muons and anti-muons (as in the η chonomic grid) can be calculated from the above by simply reversing the sign of the magnetic term in the muon and anti-muon in the g/2 factor evaluation Tables (10-4 and 10-6). This is like adding 2 x 0.000 000 337 218 . . . to the muon g/2 factor -1.001 165 912 = -1.001 165 238; and like subtracting 2 x 0.000 000 337 218 from the anti-muon g/2 factor +0.978 240 603 27 = 0.978 241 277.

Now, as before, multiply the terms and add as above: The mass ratio m/m_e for the + spin muon is 206.793 518 . . . . The mass ratio m/m_e for the + spin anti-muon is 202.081 406.

C. Orbital Masses From the η Sub-particles

Calculations are not alone for intrinsic masses on the chonomic grid. Mass ratios can and must be calculated for orbits beside intrinsic orbits—for binding orbits.

The first consideration in calculating the mass ratio of an elementary particle is to carefully examine its chonomic structure. Are there symbols on more than one state level? Or are the symbols all on one level? Is there more than one symbol on the same side of the vertical chonomic line? By inspection, determine how many different orbits are needed to bind the symbol particles together without gluons. Symbols on the same side of the vertical chonomic line (including of more than one energy state) can all be held together by means of the electric strong force—where the aether travels faster than the speed of light, the radii are made imaginary, and like charges attract. All the sub-particles in an elementary particle obey the Pauli Exclusion Principle—that is no two sub-particles in a particle have the same set of parameters—the identical state, except in briefly temporary quasi particles. Thus the sub-particles in a binding orbit can all get along together, because they are a subset of a Pauli Exclusion Principle compliant set.

In the event a central particle cannot be found, and the particle is totally symmetric, the particle acts like a two body problem, and a ½ must be put before the inertial side of the equation (see Eqn. (10-34) for a sample). In this case, one half of the charge of the sub-particles of the particle must be multiplied by the other half of the charge of the sub-particles of the particle in the electrical product portion of the balancing equation.

The mass term in the inertial side of the balancing equation is $(m/4)$ for quarton orbits, $(m/2)$ for semion orbits, and $(m/1)$ for all whole particle orbits—irrespective of mass.

The mechanic for overall orbits for oppositely charged sub-particles is similar, except it is an orbit of all the positively charged particles (bound) with all the negatively charged particles (bound). The electric term in the balancing equation is the sum of all the positive charges times the sum of the negative charges, where there is an extra minus sign because opposite charges attract.

There can be another strong force binding orbit for the charges on the other side of the vertical chonomic line. And there can be an overall binding orbit of both sides of the vertical chonomic line by the meso-electric force—where the aether travels slower than the speed of light, the radii are real, and opposite charges attract. But since the opposite particles are of opposite charges, ½ of the orbital spin is often + ½ and ½ of the orbital spin is − ½. Sometimes both sides of the orbit have like spins.

For each binding orbit, there must be an equation balancing the respective electrical force in the orbit with the inertial force in the orbit. For the Niels Bohr style calculation and approximation of the respective mass ratio associated with the orbit, the orbit is considered to be circular for particles at rest, and elliptical for particles in motion. The inertial centrifugal force is calculated from circular centripetal acceleration.

In the event a central particle cannot be found, and the particle is totally symmetric, the particle acts like a two body problem, and a ½ must be put before the inertial side of the equation (see Eqn. (10-34) for a sample). In this case, one half of the charge of the sub-particles of the particle must be multiplied by the other half of the charge of the sub-particles of the particle in the electrical product portion of the balancing equation.

The mass term in the inertial side of the balancing equation is $(m/4)$ for quarton orbits, $(m/2)$ for semion orbits, and $(m/1)$ for all whole particle orbits—irrespective of mass.

The mechanic for overall orbits for oppositely charged sub-particles is similar, except it is an orbit of all the positively charged particles (bound) with all the negatively charged particles (bound). The electric term in the balancing equation is the sum of all the positive charges times the sum of the negative charges, where there is an extra minus sign because opposite charges attract.

Let us now give three examples of the relationships between aspects of particle chonomic structures and their respective orbital balancing equations. We start with the faster than c aether mediated positively charged particle inner binding orbit for the η particle.

The total spin in the η particle chonomic structure (see Section 11 before subsection A) is 0 ℏ, which comes from the intrinsic spins on the face of the grid canceling out and the orbital spins canceling out—that is, one binding orbit has 1 ℏ spin, another binding orbit has – 1 ℏ spin, and the overall binding orbit has 1/2-1/2 = 0 ℏ spin. The negative sub particles (on the left side of the vertical chonomic line) require a faster than c binding orbit to bind like charges. The positive sub particles (on the right side of the vertical chonomic line) require a faster than c binding orbit to bind like charges. The two binding orbits require a slower than c binding orbit to bind opposite charges with the meso-electric force. Thus this particle requires three binding orbits.

The binding orbit we first wish to consider in this sample is the binding orbit of the negative particles. The + particle (electron with + ½ ℏ spin) is very slightly lighter than the – particle (electron with – ½ ℏ spin). But to all intents and purposes the – and + particles have the same mass. The binding orbit is virtually symmetric. There is no central particle in this problem. Therefore we take the two body approximation in the balancing equation. We put a ½ before the inertial side of the balancing equation. All the particles in these orbits are whole particles, so m is divided by 1. Both electrons have -1e/1 charge. We then have the following sample balancing equation for like charged sub-particles in the η particle:

$$\frac{1}{2}(m/1)v_0^2/r = (-e/1)^2 \Big/ 4\pi\varepsilon_0 \alpha^{(n/b)+1} r^2 \qquad (10\text{-}34)$$

The binding orbit we next wish to consider in this sample is the binding orbit of the positive particles. Like the negative particles on the left of the vertical line, the positively charged particles on the right of the vertical line are almost precisely the same absolute value mass. All the particles in these orbits are whole particles, so m is divided by 1. The positrons have +1e/1 charge. We then have the following sample balancing equation for like charged positive sub-particles in the η particle:

$$\tfrac{1}{2}\,1(m/1)v_0^2 / r = (e/1)^2 \big/ 4\pi\varepsilon_0 \alpha^{(N/B)+1} r^2 \,. \tag{10-35}$$

We wish now to consider the overall binding orbit of the η particle. Because of the symmetry of the negative and positive charged sub-particles, the orbit is again balanced, and requires another ½ on the inertial side of the balancing equation. We therefore use the two body approximation in this case, with whole particles and positive and negative terms in the electric term. (The negative particles have an extra negative sign because opposite charges attract in this case.)

$$\tfrac{1}{2}\,1(m/1)v_0{}^2/r = (--2e)(2e)/4\pi\varepsilon_0 \alpha^{(n/b)+1} r^2 \,. \tag{10-36}$$

Eqn. (10-34) reduces to:

$$mv_0{}^2 = 2\hbar c/\alpha^{n/b} r \tag{10-37}$$

Eqn. (10-35) reduces to the same.

Eqn. (10-36) reduces to:

$$mv_0{}^2 = 8\hbar c/\alpha^{n/b} r. \tag{10-38}$$

In Eqns. (10-37) and (10-38) we substitute for r the spin relation postulate 5 $r = n\hbar / b^2 mc$. The results are:

$v_0{}^2 = 2b^2/n\alpha^{n/b}\,c^2$ for the inner binding orbits and $\tag{10-39}$

$v_0{}^2 = 8b^2/n\alpha^{n/b}\,c^2$ for the outer binding orbit. $\tag{10-40}$

Because the outer binding orbit has ½ - ½ = 0 spin, The EF for the orbit equals 1. That is all we need, except the g/2 factors (explained above), to calculate the mass ratios for each orbit, and the total mass of the particle, and the net mass of the particle.

Now let us calculate the sub-masses for the three binding orbits in the η particle. Putting it all together we have:

Table 10-8

Full particle approximation

Sub-particle	b	n	polynomial	EF	g/2 factor	sub-mass ratio m/m_e	predicted m MeV	measured [7] m MeV
- Spin Muon	1	1	$b^2/n\alpha^{n/b}$	3/2	-1.001 165	206.793		
+Spin Muon	1	1	$b^2/n\alpha^{n/b}$	3/2	-1.001 165	206.793		
-Sp Anti-mu.	1	1	$b^2/n\alpha^{n/b}$	3/2	+0.978 165	-202.065		
+Sp Anti-mu.	1	1	$b^2/n\alpha^{n/b}$	3/2	+0.978 241	-202.081		
Orbit 1	1	1	$2b^2/n\alpha^{n/b}$	3/2	-1.001 159	412.584		
Orbit 2	1	1	$2b^2/n\alpha^{n/b}$	3/2	0.978 234	-402.159		
Orbit 3	1	1	$8b^2/n\alpha^{n/b}$	1	-0.978 234	1072.426		
Total						1092.290	558.159	547.853(24)

Outer binding orbit approximation

Sub-particle	b	n	polynomial	EF	g/2 factor	sub-mass ratio m/m_e	predicted m MeV	measured [7] m MeV
Orbit 3	1	1	$8b^2/n\alpha^{n/b}$	1	0.978 234	1072.426	548.008	547.853(24)

The outer binding orbit approximation is far closer to the measured value than the total of all mass contributions, because the particle black hole obscures all masses and orbits other than the outer binding orbit. This is good to know in our future work to save us a lot of effort, time, and paper.

As a recap, in going from the polynomials to mass ratio factors, in general, the polynomials (unitless) have to be multiplied by the kinetic-potential energy factor (unitless) times the g/2 factor (unitless) to arrive at the unitless mass ratio m/m_e. (Don't forget to add 1 for the mass ratio of the electron.) To arrive at the predicted particle mass in MeV, we multiply the resultant mass ratio by the mass of the electron in MeV (0.510 998928).

Many overall particles have 3/2 energy factors, because the kinetic energy adds to the positive potential energy of the system. But in some particle systems, the kinetic energy cancels out in the particle system, and the resultant potential energy factor is 1.

The first thing to know about g/2 factors is that they include the influence of all the forces active in the particle. The electrical medium in space has a preponderance of excess negative charged electrons. Electrons are the test charges in electrical fields. Negatively charged particles act one way in that field. Positively charged particles act another way in the field. The negatively charged particles do not have the meso-electric term (where opposite charges attract) in their g/2 factors. The positively charged particles do have the meso-electric term in their g/2 factors. The meso-electric term subtracts more and more from higher and higher energy state g/2 factors, and makes a measurable difference in 1 state, making oppositely charged particles not simple charged conjugates.

Once the exponential polynomial is multiplied by the energy factor and the g/2 factor, there is still one step that must be made before summing all the sub-mass ratios in a table: The number you have arrived at is a mass ratio factor, not the total mass ratio of the particle you are considering, unless your particle contains a ●. Then the mass ratio is as it is without summing sub mass ratios. For any state above relative ground state, without a ● in the system, the mass ratios of all lower states of the particle family must be added to the mass ratio factor to yield the mass ratio in question. Then all the mass ratios of all the sub-particles may be added together in a table for the mass ratio of the particle in question. This includes previously calculated intrinsic mass ratios on the face of the chonomic grid, as well as mass ratio equivalents calculated for the binding orbits.

The electrical force calculation for an orbit requires the identification of what the central particle in the orbit is. It is likely to be of an elevated state, or at least to be an n or an o particle. If a central particle can be identified, its charge is multiplied by the sum of the charges of the rest of the sub-particles in the orbit (for orbits binding like particles). This product is a part of the balancing equation (see Equation (10-41) for a sample).

$$1(m/1)v_0^2/r = (2e/1)^2/4\pi\varepsilon_0\alpha^{(N/B)+1}r^2 \qquad (10\text{-}41)$$

Notice the equation starts with a 1 for the single body approximation, not a ½ for a two body approximation.

---

[1]   Gordon L. Ziegler and Iris Irene Koch, "Prediction of the Masses of Charged Leptons," Galilean Electrodynamics **20** (6), 114-118 (2009).

[2]   Gordon L. Ziegler and Iris Irene Koch, "Prediction of the Masses of Every Particle, Step 1," **Galilean Electrodynamics**, Summer 2010, Vol. 21, SI No.3, pp. 43-49.

[3] Gordon L. Ziegler, *Electrino Physics* (http://benevolententerprises.org Book List; 11/23/2013, Xlibris LLC). (*Electrino Physics* Draft 2 is now available through Amazon.com and CreateSpace.com and Kindle.com.)

[4] Charles Kittel, Walter D. Knight, and Malvin A. Ruderman, Mechanics: Berkeley Physics Course–Volume 1 (New York: McGraw-Hill Book Company, 1965), p. 276.

Chapter 11

Redoing the Pion

A. Pion Family Calculations

The author has had calculations for the pion family since practically the beginning of his work of calculating the masses of "elementary" particles or chonstructs. However, the pion has been the hardest to get right. The authors have published physics books with the wrong covers on them because of persistent confusion over the structure of the pion and the forces acting in it and on it. The authors published their revised structure of the pion, but now are returning to their original structure of the pion, which is and was two quartons orbiting one way and two quartons orbiting the opposite way, superimposed on the first pair of orbiting quartons. The two quartons in one pair of quartons approach the two quartons in the other quarton pair at a relative velocity of 2 c. 1) Do they collide with each other and reflect, initiating a continual chatter—do they sing? 2) Or, since their relative velocity is over c, do they attract each other to fusion to anti-semions in anti-electrons? 3) Or since the quartons have intrinsic 0 spin other than by their orbits, and their net orbital spin is 0, do they act as bosons and go through each other without colliding?

The pion appears to be stable. Thus option 2) must not occur for the pion, though it may occur for the excited state the kaon. By definition, the single quartons as well as their collective spins are bosons. In their lowest state they must act as stable bosons. Thus the quartons in their orbits must go through each other as if their opposing quartons were not even there. The covers of *Electrino Physics* Draft 2 and *Advanced Electrino Physics* Draft 3 show the apparently correct structure of the pion among other particles.

According to postulate A.11. (Chapter 9) the mass of the particle term is the sort of exponential polynomial times the Energy Factor (EF) times the appropriate g/2 factor times the mass m_e. The exponential polynomial we can determine by solving for v_o^2 for the pion family member term. We first balance the strong electric force

187

on one quarton with the inertial force of its circular orbit. The pion has two orbiting pairs of quartons. We first solve for one of those pairs.

The effective mass of a quarton is one fourth of the mass of the whole particle in the outer non-relativistic frame. We use this mass of the quarton in the centrifugal force along with the $1/2$ from the two-body problem. The velocity v_0 is greater than c, and must increase when the energy increases. In the electric type force side of the equation, the charge of the quarton is $e/4$.

Each particle is a miniature mass singularity, and communicates with the outside world through powers of α (the Fine Structure Constant). The electric force expression, in the right side of Eqn. (11-2), we expect to depend on a power of $1/\alpha$. The numerator of the power of alpha n must be what makes the mass increase in the particle—namely the shells of mass from the radius r_j to $r \to \infty$, which can be totaled by taking $(b+1)$ (pairing of shells) times $b/2$ (number of pairs of shells). The denominator in the power of α should be b (the power of attenuation through j orders of mass singularity). Also, we want the power for the quarton orbits to be such that the power of α is 1 when $n = (b^2 + b)/2 = 0$. We take the power of α for pions and higher charged pion family members to be $n/b+1$. To solve this in MKSC (SI) Units, we must put $4\pi\varepsilon_0$ in the denominator on the right side of the equation. Balancing the forces, we have

$$\frac{1}{2}\frac{m}{4}\frac{v_0^2}{r} = (e/4)^2 \Big/ 4\pi\varepsilon_0\alpha^{(n/b)+1}r^2 \qquad (11\text{-}1)$$

for one orbiting quarton, where m is m_e. The $1/2$ in the equation is from the two-body nature of the problem, converting it to a one-body problem. The first $1/4$ in the equation is for the quarton non-relativistic effective mass. Some of the numeral constants in the above equation cancel out. The $e^2/4\pi\varepsilon_0\alpha$ can be factored out as $\hbar c$. One r can cancel out of the two sides of the equation. The equation then looks like the following:

$$mv_0^2 = \hbar c / 2\alpha^{n/b} r, \text{ where } v_0 > c \quad \text{or} \quad n, b > 0, \qquad (11\text{-}2)$$

where m equals m_e. We can eliminate m_e and r in this equation by combining this equation with Postulate A.6.

$$r = n^2\hbar/bm_e c \text{ (from postulate A.6. (Chapter 9))}. \qquad (11\text{-}3)$$

Combining Eq. (11-3) with Eq. (11-1), we can solve for v_0^2:

$$v_0^2 = (b/2n^2\alpha^{n/b})c^2, \quad b, n > 0, \qquad (11\text{-}4)$$

The m_e and r cancel out in the above calculation. Both quarton orbits in pion family members have Eqn. (11-4) as the solution of the velocity squared for those orbits. But we are doing the quarton pairs one pair at a time. We will double the masses of the quarton pairs next. The sort of exponential polynomial in the parentheses is the first unit less term we need to calculate the mass terms of the pion family members. We will double this term in our mass evaluation table.

B. The kinetic, potential, and total energies of the semion system

The kinetic, potential, and total energies of the semion system can be expressed as

$$\text{Energy}_{\text{total}} = \text{Energy}_{\text{kinetic}} + \text{Energy}_{\text{potential}} \qquad (11\text{-}5)$$

The potential energy of a mass term of a pion family member is

$$E_{\text{potential term}} = 2 \times (b/2n^2\alpha^{n/b})m_e c^2 \qquad (11\text{-}6)$$

Dividing the energy by c^2 we obtain the first term of a parameterized mass for the pion family member.

Mass is a volume thing, and is integrated from $r = r_j$ to $r \rightarrow \infty$ in discrete terms. The expression above in Eqn. (11-6), derived from first principles, is a term in a series of terms in a natural calculation of the mass of the pion family member. The sum of the terms is displayed in Table 11-1. See postulate A.12.

The Energy Factor (EF) for a single orbiting pair of quartons is 2 instead of 3/2. In electrons orbiting protons, the velocity v is much much less than c. In that case, the kinetic energy is ½ mv^2. But in the pion, the quartons orbit at c, and the kinetic energy is mc^2. That is a factor of two difference. The potential energy is also mc^2. The potential and kinetic energies are the same. So the EF factor in this case is 2.

The first author thought there should be no g/2-factors for the pion family members. But there must be $g/2$-factors after all for the pion family members. The strong term must be +1.0 instead of -1.0. And the most variable term is a term for the meso-electric force, not included in the electron $g/2$-factor. It should be

$$-bn\pi\alpha = -(p-1)n_{p-1}\pi\alpha. \qquad (11\text{-}7)$$

The g/2 factors for the pion family are predicted in *Advanced Electrino Physics* Draft 3.

C. Predictions of the Masses of the Quarton Pairs in the Pion Family

We now present Table 11-1 of the predicted (calculated from first principles) values of the masses of two quarton pairs in the pion family. In the Table, pions are denoted by π_1, kaons by π_2, D-ons by π_3, etc. Now we will double the mass calculations for the entire pion family member (see the exponential polynomial column in the table). Don't forget that the terms must be added to all previous terms to obtain the calculated predicted mass. Don't forget to multiply the result m/m_e by the mass of the electron in MeV 0.510 998 928 to obtain the calculated mass of the particle (before the mass superposition calculation).

Table 11-1
(not including superposition calculation)

Particle	b	n	$2 \times (b/2n^2\alpha^{n/b})$	EF	g/2 factor	term m/m_e	Calculated mass
Pion π_1	1	1	137.035 999	2	1.001 157 53	274.389 244	140.212 609
Kaon π_2	2	3	356.483 548	2	0.978 240 60	697.453 360	496.610 528
D-on π_3	3	6	1564.905 42	2	0.863 605 77	2702.922 70	1877.801 13

D. Masses of the anti-pion family

The anti-pion family g/2 factors are the charge conjugates of the pion family g/2 factors, except they do not have terms for the meso-electric force. Otherwise, the mass tables for the anti-pion family are the same as the mass tables for the pion family. We will denote the anti-pion as $-\pi_1$.

Table 11-2

Particle	b	n	$2 \times (b/2n^2\alpha^{n/b})$	EF	g/2 factor	Predicted m/m_e	Calculated mass
$-\pi_1$	1	1	137.035 999	2	-1.001 157 53	-274.389 244	-140.212 609
$-\pi_2$	2	3	356.483 548	2	-1.001 165 91	-713.798 352	-504.962 802
$-\pi_3$	3	6	1564.905 42	2	-1.001 157 65	-3133.434 06	-2106.144 24

These calculated masses are for both orbital pairs of quartons (see exponential polynomial term).

Now let us determine the active force between the oppositely spinning quarton pairs. Is it the strong nuclear force? The strong nuclear force is the force between nucleons like protons and neutrons, and pions are the mediating particle. Pions are strong gravitationally attracted entities to both nucleons, and are the go-between particles between the nucleons force wise. In nuclei, the pions have the leading important role. In the pion attached to a neutron to make it a proton, as in baryon structure in our chonomic system, is the strong nuclear force active with only one nucleon at a time (neutron or proton)? No. Pions mediate the strong nuclear force between nucleons, but a single pion is not attracted to a single neutron by the strong nuclear force.

What force then is active between the two halves of the pion? Is it the magnetic force? The magnetic force is active in this problem and adds a little to the maximum energy term a g/2 factor can have, but does not integrate well as a stick on force. According to the magnetic integration between two different levels of magnetism, the energy difference of the two levels calculates to be 0, because the magnetic field is a closed loop.

What force then is active between the two halves of the pion? The aether in the pion travels at or faster than c. So the strong electric force is active of like attracting like charges in the pion. The quartons in the pion are all positively charged. They attract each other. The integrable force in a pion is the virtual center of mass charge of one orbiting quarton pair attracting the other orbiting quartons. When the orbiting pairs coincide, the separation of the centers is not taken to be zero (which would lead to an infinite force), but r separation of one center to both the quartons in the opposite pair of quartons. The integral to be evaluated is

$$(2)\int_{\infty}^{r} \frac{e}{2} \frac{e}{4} \frac{1}{4\pi\varepsilon_0 \alpha r^2} dr = \frac{-\hbar c}{4r} + 0. \qquad (11\text{-}12)$$

This is the integral of the strong electric force between the orbiting quarton pairs. In integrating a force times dr, we obtain the energy difference between the quarton pairs from infinity separation to r separation. The (2) in the equation is for the two quartons in orbit 2 experiencing the force integration to the quarton pair 1 (we are taking both quartons together in this portion of the calculation to add to the double of the mass of the quarton orbits calculated previously); the e/2 is for the quarton pair 1 attracting quarton pair 2. The e/4 is for a single quarton charge attracted to quarton orbital pair 1 (doubled by (2)). We can combine Eqn. (11-3) with Eqn. (11-12) to eliminate the explicit r from the equation. We obtain

$$\Delta E = -bm_e c^2/4n^2 . \qquad (11\text{-}13)$$

By dividing by c^2, we obtain the mass difference to add to the mass values in Tables 11-1 and 11-2.

$$\Delta m = -bm_e/4n^2. \tag{11-14}$$

We note that, except for the m_e, the above expression is similar to the sort of exponential polynomial for the orbits. The question comes, should there be also an EF factor and a g/2 factor in this radial integration? The expression in Eqn. (11-14) is for the potential energy difference between infinity and r. But a quarton accelerated from infinity to r would have a velocity also and a kinetic energy radially (not orbitally). The velocity would be negative. And times a positive velocity, the v_r^2 would be negative, and the kinetic energy would be the same as the potential energy. So it would be correct to expect that there should be an EF factor of 2 also for the radial integration and a g/2 factor.

Table 11-3

Pion family member	b	n	$-b/4n^2$	EF	g/2	integrated mass MeV	orbital mass MeV	total mass MeV	measured mass MeV [1]
Pion π_1	1	1	-0.250 000	2	1.001 157	-0.255 795	140.212	139.956	139.570
Kaon π_2	2	3	-0.055 555	2	0.978 240	-0.055 542	496.610	496.554	493.677
D-on π_3	3	6	-0.020 833	2	0.863 605	-0.018 387	1877.801	1877.78	1869.62

Table 11-4

Pion family member	b	n	$-b/4n^2$	EF	g/2	integrated mass MeV	orbital mass MeV	total mass MeV	measured mass MeV [1]
Anti-pion	1	1	-0.250 000	2	-1.001 157	0.500 578	-140.212 609	-139.712	not
Anti-kaon	2	3	-0.055 555	2	-1.001 165	0.056 843	-504.962 802	-504.905 9	measured
Anti-D-on	3	6	-0.020 833	2	-1.001 157	0.021 316	-2106.104 48	-2106.083	yet

The calculated pion mass has three place accuracy to the measured pion mass. That is better than the original calculation. The kaon and D-on have two place accuracy compared with the measured masses.

---

[1]   J. Beringer **et al.** (Particle Data Group) PR **D86**, 010001 (2012) and 2013 update for the 2014 edition (URL: http://pdg.lbl.gov).

Chapter 12

Redoing the Neutron

The third particle type—the neutron family—is different from the other two particle types. The neutron is a baryon, and, like all baryons, has affecting the mass calculations not only a relativistic imaginary-axis massive core, but also at the same time a real-axis zero mass core—the uniton. Electron family members orbit about this massive core particle. It is similar to electrons orbiting protons in Hydrogen. But electrons orbiting a proton are easily ionized, whereas the electron orbiting the uniton in a neutron is strongly bound to the core uniton. The uniton cannot come alone. For all lower particles, an electron always accompanies the uniton. Electrons orbiting protons could have semions in higher states, but this is not normally considered. Likewise the semions in the electron orbiting the uniton in neutrons could have higher states, but we do not consider them in our study of the neutron. It is the electron states orbiting the uniton in the neutron that can have a range of values in the electron orbits in the neutron.

The neutron has an interesting g/2 factor. Neutrons have positive cores, but they do not have positive strong force terms in their g/2 factors. They have negative strong force terms. [2](Chapter 5). What can account for this? Dots may be obscured by black holes, whereas – particles in the neutron (electrons), are outside the black holes and not obscured. The +1 strong force of the dot is obscured. The -1 strong force of the – particle (electron orbiting the dot uniton) in the g/2 factor is not obscured. Therefore, to all appearances, the negative strong force replaces the positive strong force in the neutron. The same goes for the electric and magnetic terms.

The anti-neutron family has a + 1 strong force term from the orbiting positron (see above), no meso-electric term, and the charge conjugate of all the other terms as the neutron family member.

A binding orbit uses the g/2 factor of its central particle or the highest mass ratio sub-particle. Calculations are best if the energy factor and the g/2 factor are multiplied separately with each polynomial, and the mass ratios summed up.

Just as each previous particle family type has a different spin relation, the electron orbits of the uniton have a different spin relation. With the neutron family orbits, we will use B and N for the whole particle electron orbiting the uniton, to differentiate it from b and n in the electron family orbits (also used in these calculations). The total neutron spin is $mv_o r$, but the observable spin on the event horizon of the black hole is only mcr. We have to go by the observable spin. The neutron observable spin relation is:

$$m_e cr = N\hbar/B \tag{12-1}$$

$$r = N\hbar/Bm_e c \tag{12-2}$$

Let us balance the force equation for the neutron family. Instead of the balancing of the electric force and the inertial force in Eqn. (12-3) starting with a 1/2, as in Eqns. (10-34), (10-35), and (10-36), because of the two-body nature of those problems, Eqn. (12-3) starts with a '1', because of the single body nature of the neutron family problem. In this problem, the mass is $m/1$ instead of $m/2$, because in the main electron family orbits about the uniton, we are dealing with electrons as whole particles, not half particles.

$$1(m/1)v_o^2/r = (e/1)^2/4\pi\varepsilon_0\alpha^{(N/B)+1}r^2 \tag{12-3}$$

Using the techniques under Eqn. (10-36), this reduces to

$$mv_o^2 = \hbar c/\alpha^{N/B}r \quad . \tag{12-4}$$

Combining this with Eq. (9-5), we obtain

$$mv_o^2 = \hbar c Bmc/N\hbar\alpha^{N/B} \tag{12-5}$$

$$v_{on}^2 = B/N\alpha^{N/B} c^2 \tag{12-6}$$

This is the orbital velocity squared for the neutron orbit of the electron family member. We have to combine that with the orbital velocity squared for the electron family member v_{oe}^2, which is

196

solved by combining the electron spin relation substituting b for B and n for N in Eqn. (12-6), because the neutron space charge limits the spin relation for the intrinsic electron family members in the neutron, and is

$$v_{oe}^2 = b/n\alpha^{n/b} c^2 \qquad (12\text{-}7)$$

There is a region where v_{on} relative to v_{oe} is faster, and a region where it is slower, but the average of the absolute value v_{on} and average absolute value v_{oe} are at right angles to each other and can be added by squaring them. The process is clarified in Eqn. (12-8).

$$v_{Tn}^2 = [(B/N\alpha^{N/B}) + (b/n\alpha^{n/b})] c^2 \qquad (12\text{-}8)$$

By multiplying the quantity in Eq. (12-8) by m_e we obtain the potential energy of the neutron family system. By multiplying the potential energy of the neutron family members by the EF $3/2$, we obtain the total energy including the kinetic energy in the neutron. By dividing by c^2 we obtain the mass m of the neutron family member. Each particle has a $g/2$-factor. [1, 2] The result is in Eqn. (12-9).

$$m_{particle} = [(B_i/N\alpha^{N/B} \text{ EF } g_i/2) + (b_j/n\alpha^{n/b} \text{ EF } g_j/2)] m_e \qquad (12\text{-}9)$$

Unitons are different from other whole-body systems. There are no elevated states of unitons. They are always only at state 2. The only elevated states associated with unitons are with orbiting particle systems surrounding the unitons. This feature of unitons apparently is reflected in the property that uniton systems have only one shell of mass. The sum of the velocity and thus mass components of the electron family member and the neutron system is not compounded by layers of mass shells. That typical stage of calculations will be left out of baryon calculations.

We see the mass of the neutron is a combination of two calculable terms. The N's and the n's are calculable from the B's and the b's. To calculate this in general, we must have a definition of n, b, and j: (See Table 12-1.). The first three n and b

197

are tested. Higher n and b are calculated. We expect both n and b to increase with j. We expect $n_j - (n_{j-1})$ to be b_j.

Table 12-1

j	0	1	2	3	4	5	6
n	0	1	3	6	10	15	21
b	0	1	2	3	4	5	6

For all neutron family members, the orbital spin is -1. The B for all neutron family members is 2. On the other hand, the minimum b and n are 0. The neutron family members all have J = ½ and parity +. They all have unitons for core particles. They all have an electron family member with $\hbar/2$ intrinsic spin orbiting around the uniton with $\hbar$ orbital spin. The only things that differentiate the neutron family members are the energy states of the particles. Yet all the observed particles with these properties have different, seemingly unrelated, traditional names. Those observed so far are n, Λ, Σ^0, and Λ_b^0. To show the neutron related nature of those particles, we shall name those same particles $n = n_1$, $\Lambda = n_2$, $\Sigma^0 = n_3$, and $\Lambda_b^0 = n_4$, etc.

The g/2 terms are simplified also. There is only one g/2 factor for the b terms—at b = 2 (Λ g/2 factor, see [2] Chapter 5). For the intrinsic states of the semion orbits in the orbiting electron about the uniton, see the electron family g/2 factors in [2] Chapter 5.

With the neutron family members, there is too much information to put in one table. We will divide the information into three tables.

Particle		b	n	$b/n\alpha^{n/b}$	B	N	$B/N\alpha^{N/B}$
n	n_1	0	0	1.000 000 000	2	3	1,069.450 645
Λ	n_2	1	1	137.035 999 7	2	3	1,069.450 645
Σ^0	n_3	2	3	1,069.450 645	2	3	1,069.450 645
Λ_b^0	n_4	3	6	9,389.432 523	2	3	1,069.450 645

Table 11-2 Neutron family parameters

In the next table, the g/2 factor to be utilized for B = 2 and N = 3 is the one where the meso-electric factor is -2 x 3πα, or the g/2 factor for the Λ particle.

Parti-cle	b_i	n_i	$g_i/2$	B_j	N_j	$g_j/2$
n	0	0	-1.000 000 000	2	3	-1.138 716 794
Λ	1	1	-1.001 165 912	2	3	-1.138 716 794
Σ^0	2	3	-1.001 157 653	2	3	-1.138 716 794
Λ_b^0	3	6	-1.001 165 744	2	3	-1.138 716 794

Table 12-3 System g/2 factors

Particle	Predicted Mass Ratio (m/m_e)	Measured Mass Ratio [3] (m/m_e)
n	1,828.202 115	1,838.683 66
Λ	2,032.495 772	2,183.337
Σ^0	2,288.490 108	2,333.9
Λ_b^0	10,618.179 61	10,996

Table 12-4

All calculated values, except for the Λ, are two place accuracy to the measured values. The calculated mass ratios of each neutron family member are a little on the low side compared to the measured values. This is to be expected because we are calculating only circular orbits and neglecting relativistic effects. But actually, our calculated values are a pretty good fit with the measured values. When our calculated values go up by small steps, the measured values go up by small steps. And when our calculated value goes up by a large step, the measured value goes up by a large step—with about the same degree of precision.

[1]   Gordon L. Ziegler and Iris Irene Koch, *Advanced Electrino Physics* Draft 2. Draft 3 now available at Amazon.com.

[2]   Gordon L. Ziegler and Iris Irene Koch, "Prediction of the Masses of Every Particle, Step 1," **Galilean Electrodynamics**, Summer 2010, Vol. 21, SI No.3, pp. 43-49.

[3]   J. Beringer *et al* (Particle Data Group) PR **D86**, 010001 (2012) and 2013 partial update for the 2014 edition (URL: http://pdg.lbl.gov).

GAUGE AND HIGGS BOSONS

γ

$I(J^{PC}) = 0,1(1^{--})$
Mass $m < 1 \times 10^{-18}$ eV
Charge $q < 5 \times 10^{-30}$ e
Mean life τ_γ = Stable

γ
0

2    $\bullet \mid \bullet$
1    $\mid$
0    $\mid$

$\underline{1 \mid 1}$
$\mid$

The gamma or photon is the one particle that cannot be constructed of electron, pion, or neutron family members. [1] (p. 43). This is because the fundamental whole particle the neutron is not elementary. It is composed with a zero spin nearly point charge uniton and an electron orbiting it. It is a piece of the neutron—the uniton—and the anti-uniton that compose the photon. The uniton has imaginary mass $-i2.17644(11) \times 10^{-08}$ kg. [2] (Chapter 6, Eqn. (6-17). The anti-uniton has minus that quantity, or $+i$ times the value. The net charge mass of the photon is 0, so it travels linearly at the speed of light c. [2] (Chapter 6, Eqn. (6-14). But the photon probably cannot be summed at once. It has the uniton and the anti-uniton on opposite sides of a black hole. Only one can be seen at any one time. The photon has charge and mass oscillations as it travels along the light path axis. To be sure, the photon has time average zero mass and charge. But the photon has energy by virtue of its oscillation: $E = h\nu$. The photon has orbital spin of $\pm 1\hbar$ and travels at the speed of light.

Let us try the mass calculating procedure developed in previous chapters, to see if it gives us the right value for the photon.

The uniton and anti-uniton are held together by the meso-electric force in slightly slower than light non-relativistic calculations. The balancing equation for the uniton is

$$\tfrac{1}{2}(m_e/1)v_0^2/r = (-e)(e)/4\pi\varepsilon_e\alpha^{n/b+1}r^2 \tag{13-1}$$

The $e^2/4\pi\varepsilon_0\alpha$ can be factored out as $\hbar c$. One r can cancel out of the two sides of the equation. The equation then looks like the following:

$$m_e v_0^2 = 2\hbar c/\alpha^{n/b}r. \tag{13-2}$$

The 2 in the above equation comes from the 2 in the $\tfrac{1}{2}$ for the two body problem. The spin relation for the dot (●) uniton is

$$r = n\hbar/bm_e c. \quad [1] \text{ p. 48}). \tag{13-3}$$

Combining Eqn. (13-2) with Eqn. (13-3), we can solve for v_0^2:

$$v_0^2 = (2b/n\alpha^{n/b})c^2. \tag{13-4}$$

The anti-uniton has $v_0^2 = -(2b/n\alpha^{n/b})c^2$. Thus we see again that the mass of the photon equals zero.

g $I(J^P) = 0(1^-)$
or gluon   Mass $m = 0$
                        SU(3) color octet

There are no gluons in the electrino system.

W   J = 1
                       Charge = ±1e
          Mass m = 80.385 ± 0.015 GeV [3] (measured at
velocities close to the speed of light)
                       W
                       ?
      5 | o
      4 |
      3 |
      2 |
      1   - | -
      0 |

         1-1 | -1
           |

The W and Z particles have more mass than any other particles in
this three volume set *Predicting the Masses*. They either have an
extreme chonomic structure as displayed here, or their extremely
high mass is due to reaction velocities close to the speed of light.
The W particle has listed with it [3] a very large momentum. To get
a more reliable mass estimate of these particles, they should first be
stopped and then measured for mass.

Z   J = 1
               Charge = 0
               Mass m = 91.1876 ± 0.0021 GeV [3]
      5   + | +
      4 |
      3 |
      2 |
      1 |
      0 |

         1-1 | -1
           |

Higgs Bosons

A high energy boson has recently been discovered. The authors believe it will be found to not have Higgs Boson properties—that is, it will not determine the masses of all other particles. In the Electrino Model of Physics, each particle determines its own masses from first principles by the radii of their orbits. There is no provision for Higgs Bosons in this Chonomic System.

----

[1] Gordon L. Ziegler and Iris Irene Koch, "Prediction of the Masses of Every Particle, Step 1," **Galilean Electrinodynamics**, Summer 2010, Vol. 21, SI No. 3, pp. 43-49.

[2] Gordon L. Ziegler, *Electrino Physics* Draft 2, available from Amazon.com, CreateSpace.com and Kindle.

[3] J. Beringer *et al* (Particle Data Group) PR **D86**, 010001 (2012) and 2013 partial update for the 2014 edition (URL: http://pdg.lbl.gov).

Chapter 14

FRACTONS

QUARKS

No formulation of quarks in the electrino system.

ELECTRINOS

Except for unitons, electrinos cannot be shown as symbols on the chonomic grid. The chonomic grid is for whole particles. Except for unitons, electrinos are fractons, or fractional charged particles; and unitons act like fractons. They never come alone. They always come with another particle—an anti-uniton in photons, or electrons in neutrons, or + echons in Σ particles.

"In addition to the electric self potential mass, the uniton or any electrino has kinetic mass in the electrino relative rest frame. The aether field is not static. Every portion of charge in the uniton is traveling at speed c relative to the aether. Therefore m_q or M_q has kinetic energy also. If m_q had a small velocity relative to the aether, its kinetic energy E_{kin} would be $\frac{1}{2}m_q v^2$. But since $v^2 = -c^2$, we take the relativistic form $E_{kin} = -m_q c^2$. The total fundamental mass of the uniton is $M_q + (-M_q) = 0$. The total absolute value mass is $|M_q| + |-M_q| = 2M_q = M_0$, which is the imaginary Planck mass, composed simply of the following constants:

$$M_0 = -i\left(\frac{\hbar c}{G}\right)^{1/2}. \qquad (14\text{-}1)$$

Numerically it is

$$M_0 \approx -i\ 2.176\ 44(11)\ x\ 10^{-08}\,kg. \text{[1]} \qquad (14\text{-}2)$$

$$R_0 = \frac{2GM_q}{-c^2} = i\left(\frac{\hbar G}{c^3}\right)^{1/2} \approx i\ 1.616\ 252(81)\ x\ 10^{-35}\,m. \qquad (14\text{-}3)$$

"The physical size of the uniton is very small—essentially a point charge. But it is imaginary in radius. The mass in the relative rest frame is very large on particle scales. But it is minus imaginary. Essentially the radius of the uniton has been relativistically contracted and the mass relativistically increased. The circumferences of the uniton, however, are not in the direction of aether motion. We might think they are not contracted. But they are. The circumference is 2π times the relativistic imaginary radius. The relativistic particles are relativistic throughout.

"Semions and quartons (1/2 and 1/4 charges) are also of interest to us. From equations parallel to [1] Equations (6-7) through (6-18) we see that, while the fundamental masses of the particles are all zero, the absolute value masses and radii are:

$$m_{quarton} = 1/4\ m_{uniton} \approx -i\ 5.441\ 1(03)\ x\ 10^{-09}\,kg. \qquad (14\text{-}4)$$

$$r_{quarton} = 1/4\ r_{uniton} \approx i\ 4.040\ 63(21)\ x\ 10^{-36}\,m. \qquad (14\text{-}5)$$

$$m_{semion} = 1/2\ m_{uniton} \approx -i\ 1.088\ 22(06)\ x\ 10^{-08}\ kg. \qquad (14\text{-}6)$$

$$r_{semion} = 1/2\ r_{uniton} \approx i\ 8.081\ 26(41)\ x\ 10^{-36}\,m." \text{[1] (Chapter 6, Section}$$
II E). $\qquad\qquad (14\text{-}7)$

The above electrino masses are all from the relativistic frame, and are precise. The non-relativistic electrino masses are not unique, but depend on the particle they are found in. For instance, the non-relativistic semion mass equals $(1/2)m_e$ or $(1/2)m_\mu$ or

$(1/2)m_\tau$ or etc. The non-relativistic quarton mass equals $(1/4)m_\pi$ or $(1/4)m_K$ or $(1/4)m_D$ or etc.

---

[1] Gordon L. Ziegler and Iris Irene Koch, *Electrino Physics* Draft 2. Available at amazon.com. createspace.com, and kindle.com.

Chapter 15

Leptons

. e $J = \frac{1}{2}$

Mass m = 0.510998928 ± 0.000000011 MeV
  = (548.57990946 ± 0.00000022) x 10^{-6} u
$|m_{e+} - m_{e-}|/m < 8$ x 10^{-9}, CL = 90%
$|q_{e^+} + q_{e^-}|/e < 4 x 10^{-8}$
Magnetic moment anomaly
  (g-2)/2 = (1159.65218076 ± 0.00000027
$\left(g_{e^+} - g_{e^-}\right)/ g_{average} = \left(-0.5 \pm 2.1\right) x 10^{-12}$
Electric dipole moment d < 10.5 x 10^{-28} ecm, CL = 90%
Mean life $\tau > 4.6$ x 10^{26} yr, CL = 90%

```
            e
  0.510  998  928
     2    |
     1    |
     0   -|

      0 | -½
        |
```

   The electron mass is the one particle mass that has to be
input from the measured value. $m_e = 0.510\ 998\ 928(11)$ MeV [1]

. μ $J = \frac{1}{2}$

Mass m = 105.6583715 ± 0.0000035 MeV
= 0.1134289267 ± 0.0000000029 u
Mean life τ = (2.1969811 ± 0.0000022) x 10^{-6} s
Magnetic moment anomaly (g-2)/2 = 11659209 ± 6) x 10^{-10}

μ
105.671 336 5
2 ⊥
1 ⊥
0 |

0|-½
|

The ratio m_{μ} / m_{e} is already calculated in [2], [3], [4] and [5], but calculated by an incorrect formula. Let us update those calculations by the correct formula.

$$m = \left\{ \left(\frac{b_j^2}{n_j \alpha^{n_j / b_j}} \right) EF_j (g/2)_j + \left(\frac{b_{j-1}^2}{n_{j-1} \alpha^{n_{j-1}/bj-1}} \right) EF_{j-1} (g/2)_{j-1} + ...1 \right\}$$

m_e. (15-1)

We already evaluated this formula for the muon:

Particle	b	n	polynomial	EF	g/2 factor	mass factor m/m_e
Electron	0	0	included	included	included	1.000 000
Muon	1	1	1/α	3/2	-1.001 165 912	205.793 656

| | Total Muon | 206.793 656 |

Table 15-1

Let us evaluate formula (15-1) for the tauon:

Particle	b	n	polynomial	EF	g/2 factor	mass factor m/m_e
Electron	0	0	included	included	included	1.000 000
Muon	1	1	$1/\alpha$	3/2	-1.001 165 912	205.793 656
Tauon	2	3	$4/3\alpha^{3/2}$	3/2	-1.001 157 653	3,212.066 092

Total Tauon   3,418.859 748

The tauon mass in MeV can be derived by multiplying the total tauon m/m_e by $m_e = 0.510\,998\,928$ MeV. The tauon mass in Mev is 1747.03, which is in the calculated mass position in the chonomic structure below. Parameters l (elliptical) and m (tilt or magnetic) are not included in any of the calculations in this book—a potential source of errors in the calculations.

. τ $J = \frac{1}{2}$

Mass $m_\tau = 1776.82 \pm 0.16$ MeV
    τ
1747.03 MeV
2    $-|$
1    $|$
0    $|$

0 | $-\frac{1}{2}$
$|$

---

[1] J. Beringer *et al* (Particle Data Group) PR **D86**, 010001 (2012) and 2013 partial update for the 2014 edition (URL: http://pdg.lbl.gov).

[2]   Gordon L. Ziegler and Iris Irene Koch, "Prediction of the Masses of Charged Leptons," **Galilean Electrodynamics 20** (6), 114-118 (2009).

[3]   Gordon L. Ziegler and Iris Irene Koch, "Prediction of the Masses of Every Particle, Step 1," **Galilean Electrodynamics**, Summer 2010, Vol. 21, SI No.3, pp. 43-49.

[4] Gordon L. Ziegler, *Electrino Physics,* 11/23/2013, Xlibris LLC; (http://benevolententerprises.org Book List).   *Electrino Physics* Draft 2 is now available at amazon.com, createspace.com, and kindle.com.

[5]   Gordon L. Ziegler and Iris Irene Koch, *Advanced Electrino Physics* Draft 2. *Advanced Electrino Physics* Draft 3 will be available in a few days from 02/20/2015. Check and see if it is available yet.

Chapter 16

Neutrinos

Neutrinos have given particle theorists some degree of problems. At first the neutrinos were thought to have zero mass and travel at the speed of light. But then the work at the Super-Kamiokanda in Japan [1], [2], and [3] disproved the zero neutrino mass premise of the Standard Model. Not long ago neutrinos were thought to be clocked faster than the speed of light. [4]. Thousands of scientists now dissent from Einstein's Special Theory of Relativity [5], particularly his conclusion that nothing could travel faster than the speed of light, and that there is no aether. Aether theories of relativity now abound. But some Standard Model loyalists have seriously objected to the neutrino clocking experimental results, attributing them to various experimental errors. They are not ready to abandon Einstein's Special Theory of Relativity—a pillar in the Standard Model.

The postulate that nothing can go faster than the speed of light does not work in the authors' unified field theory [6]. The only possible way in this theory to unite all the forces is to allow some classes of particles to travel faster than the speed of light—especially orbitally. When the speed of the particles is slightly faster than the speed of light relative to the aether, then their radii are imaginary. And if two of them are multiplied together, it introduces an extra − sign into the force. Like particles then attract instead of repel. This is a valuable result which simply unites the forces, which could not be had if nothing traveled faster than the speed of light. There once was a theory of tachyons—some particles born faster than the speed of light. But Standard Model loyalists do not give credence to that theory. The authors take their stand with the idea that some particles like neutrinos do travel faster than the speed of light.

We now have a new tool to attack this mystery—a road map on how to calculate masses of particles from first principles. Let us try the method of calculation in Chapter 10 and Chapter 11 on the electron neutrino. The given chonomic structure of the electron neutrino from Chapter 8 is

ν_e

$$J = \frac{1}{2}$$
$$\text{Mass m } <0i \text{ MeV}$$

ν_e

-i0.525 eV   (calculated in this chapter)

```
2      |
1      |о
0     -|
```

```
1 | ½
  |
```

The intrinsic mass ratios on the face of chonomic grid are 1.000 000 000 for the − echon—the electron; and 273.132 04 for the o echon—the pion [7]. The masses in MeV of the particles are 0.510998928 MeV for the electron, and 139.570 179 6 MeV for the pion. The two particles are in orbit around each other with orbital spin 1 ħ and traveling at right angles linearly slightly faster than the speed of light. The two particles have opposite charges, and they attract each other with the slower than the speed of light meso-electric force. Thus we are led to believe that they orbit around each other just slightly slower than the speed of light.

But there is another speed of light we have to be concerned with with neutrinos—their linear translational velocity. Relative mass increase for the neutrinos places them in the minus imaginary mass portion of the relative mass increase curve, which we take to be as valid as the relative mass increase curve slower than the speed of light. The super luminal portion of the curve is not a waste as Einstein would have it.

This orbit has a central particle—the pion, which is much more massive than the electron. Thus we have a one body problem here, not a two body problem. Thus the balancing equation below must begin with a 1 not a 1/2.

But the electron orbits around the pion much as the electron orbits around the uniton in the neutron. Like the neutron, the neutrino has two mass terms which must be added. One, for the electron portion of the neutrino (the intrinsic mass of the electron), must have the constants b and n equal to 1 and 1 for all neutrino

213

states. The other mass value (the electron orbiting the pion), must have different B and N states starting at 1 and 1 and going up from there. Providence has ruled that the mass of all the particles in the Universe except the electron should be calculable without definition.

The neutrino problem is most like the Neils Bohr's electron orbiting a proton problem, except we have a Fine Structure α of the meso-electric force. It is not like the intrinsic mass calculations for electrons or pions. Those are different states of simple electrons or pions, not electrons orbiting about a positive particle. It is not like electrons orbiting around protons as in Niels Bohr's calculations. Those are with the Coulomb electric force, not with the meso-electric force as in the neutrino problem. But the neutron problem is also with electrons orbiting positive particles with the meso electric force. Therefore we choose the neutron problem as the most similar to the neutrino problem for the exponential polynomial. We therefore assume that we can use the neutron spin relation (tested) for the electron neutrino problem. From Chapter 9, Postulate A.7. we obtain

$$r_1 = N\hbar/Bm_1c. \tag{16-1}$$

Let us balance the force equation for the neutrino family. Instead of the balancing of the electric force and the inertial force starting with a 1/2, because of the two-body nature of those problems, Eqn. (15-2) starts with a '1', because of the single body nature of the neutrino family problem. In this problem (the pion portion of the electron neutrino), the mass is $m/1$ instead of $m/2$, because we are dealing with electrons as whole particles, not half particles. Different than mass particles, neutrinos travel linearly slightly faster than the speed of light. Therefore their masses we want to solve for are minus or plus imaginary values. We want to trust the balancing equation to solve for the correct values of the imaginary masses. But the radius depends on the direction measured. The radius in the translational direction is ir_2, but the radius of the orbit is the real r_1. We have to be careful of that distinction in the balancing equation below. Also, we cannot cancel the $-im$ with a real m. This means that we do not yet have sufficient number of independent equations to solve the increasing number of variables in this system even with the appropriate spin postulate for this problem.

$$1(-im_2/1)v_o{}^2/r = (-e)(+e)/4\pi\varepsilon_0\alpha^{(N/B)+1}r^2 , \qquad (16\text{-}2)$$

for the oppositely charged electron and pion held together by the meso electric force. Using the techniques in Chapter 10, this reduces to

$$-im_2v_o{}^2 = -\hbar c/\alpha^{N/B}r \quad . \qquad (16\text{-}3)$$

Combining this with Eq. (16-1), we obtain

$$-im_2v_o{}^2 = -\hbar cBmc/N\hbar\alpha^{N/B} \qquad (16\text{-}4)$$

Simplifying we have

$$-im_2v_o{}^2 = -Bm/N\alpha^{N/B} c^2 \qquad (16\text{-}5)$$

We cannot cancel a m with an im_2. We therefore need another independent equation to solve for $-im_2$. The relative mass increase formula for m is an independent equation to help us in this system:

$$-im_2 = (1 - v_2{}^2/c^2)^{-1/2}m \qquad (16\text{-}6)$$

Let us substitute for $-im_2$ in Eqn. (16-5) and solve for $v_o{}^2$:

$$(1 - v_2{}^2/c^2)^{-1/2} v_o{}^2 = -B/N\alpha^{N/B} c^2 \qquad (16\text{-}7)$$

$$v_{op}{}^2 = -B/N\ \alpha^{N/B} (1 - v_2{}^2/c^2)^{1/2} c^2 \qquad (16\text{-}8)$$

This is the orbital velocity squared for the neutrino orbit of the electron family member. We have to combine that with the orbital velocity squared for the electron family member $v_{oe}{}^2$, which is solved by combining the electron spin relation substituting b for B and n for N in Eqn. (12-6), because the neutron space charge limits the spin relation for the intrinsic electron family members in the neutron, and is

$$v_{oe}^2 = b/n\alpha^{n/b} c^2 \qquad (16\text{-}9)$$

There is a region where v_{on} relative to v_{oe} is faster, and a region where it is slower, but the average of the absolute value v_{on} and average absolute value v_{oe} are at right angles to each other and can be added by squaring them. The process is clarified in Eqn. (16-10).

$$v_{Tn}^2 = [(B/N\alpha^{N/B}) + (b/n\alpha^{n/b})] c^2 \qquad (16\text{-}10)$$

By multiplying the quantity in Eq. (16-10) by m_e we obtain the potential energy of the neutrino family system. By multiplying the potential energy of the neutrino family members by -3/2, we obtain the maximum absolute value total energy including the kinetic energy in the neutrino. By dividing by c^2 we obtain the mass m of the neutrino family member. Each particle has a $g/2$-factor. [7] The result is in Eqn. (16-11).

$$m_{neutrino} = [-(B_i/N\alpha^{N/B} EF_i\, g_i/2) + (b_j^2/n\alpha^{n/b} EF_j\, g_j/2)] (1 - v_2^2/c^2)^{\frac{1}{2}}\, m_e$$
(16-11)

We see the mass of the neutrino is a combination of two calculable terms. The N's and the n's are calculable from the B's and the b's. To calculate this in general, we must have a definition of n, b, and j: (See Table 16-1.). The first three n and b are tested. Higher n and b are calculated. We expect both n and b to increase with j. We expect $n_j - (n_{j-1})$ to be b_j. We leave out 0 and 0 for all particles other than electrons and electron neutrinos.

Table 16-1

j	0	1	2	3	4	5	6
n	0	1	3	6	10	15	21
b	0	1	2	3	4	5	6

For all neutrino family members listed, the orbital spin is +1. The neutrino family members all have $J = \frac{1}{2}$. They all have pions for core particles. They all have an electron family member with $-\hbar/2$ intrinsic spin orbiting around the pion with $\hbar$ orbital spin. The only things that differentiate the neutrino family members are the energy states of the particles. The minimum B for N for electrons orbiting the pion is 0 and 0. The only b and n for the exponential polynomial we will consider for the electron are 1 and 1, and for the g/2 factor are 0 and 0.

The g/2 terms are simplified also. There is only one g/2 factor for the b terms—at b = 0 (See [7] Chapter 5).

In the table below is v_2 in a calculation. We allude to part of an article by Robert Evans to document our inference of v_2. [4] In that article is reported measurements of neutrinos taking 60 nanoseconds less time to travel the distance light would travel in 2.4 thousandths of a second. We calculate from that information v_2:

$$v_2 = 2.4\text{E-}03/(2.4\text{E-}03 - 60\text{E-}09) \, c = 1.000025001 \, c. \qquad (16\text{-}12)$$

$$v_2^2/c^2 = 1.000050002 \qquad (16\text{-}13)$$

$$(1 - v_2^2/c^2) = (1.0 - 1.000050002) = -0.00005000189. \qquad (16\text{-}14)$$

$$(1 - v_2^2/c^2)^{\frac{1}{2}} = i0.007\,071\,201\,454. \qquad (16\text{-}15)$$

where v_2 may be different for electron neutrinos, muon neutrinos, and tauon neutrinos. The authors assume that it was electron neutrinos that were clocked in the experiment, not muon neutrinos or tauon neutrinos. We take the clocked velocity of the neutrinos as the value of v_2 of the electron neutrinos (slightly faster than the speed of light).

It is not likely that v_2 is the same for all neutrinos. But in the off chance that it is, we only have to calculate by the B and N values for the different based neutrinos to not only calculate the electron based neutrino mass, but also the muon and tauon neutrino masses. The appropriate values of B and N are in the second through fourth columns of Table 16-1.

All we need now are the appropriate EF factors (Energy Factors) and the appropriate g/2 factors to solve for the mass of the electron neutrino and guess the masses of the higher based observed kinds of neutrinos (in the off chance that v_2 is the same for all neutrinos). The spins are not canceled out in the neutrinos; but the potential energies are negative; therefore the EF is -1/2 or -3/2 for each neutrino, depending if it is + or − spin. To be in harmony with data reporting conventions, however, we take the − intrinsic spin and orbital spin equal to +1 and the total neutrino spin as +½ and the neutrino EF to be − ½.

With the neutrino family members, there is too much information to put in one table. We will divide the information into three tables for the neutrinos and three tables for the anti-neutrinos.

Particle	B	N	$- B/N\alpha^{N/B}$	b	n	$b/n\alpha^{n/b}$	$(1-v_2^2/c^2)^{½}$
v_e	1	1	-137.035 999	1	1	137.035 999	i0.007 071
v_μ	2	3	-1,069.450 64	1	1	137.035 999	i0.007 071
v_τ	3	6	-9,389.432 52	1	1	137.035 999	i0.007 071

Table 16-2 Neutrino family parameters

In the next table, the g/2 factor to be utilized for B = 1 and N = 1 is the one where the meso-electric factor is $-(b-1)n\pi\alpha$, or the g/2 factor for the pion particle. The g/2 factor for the second term is the muon value.

Particle	B_i	N_i	$g_i/2$	b_j	n_j	$g_j/2$	EF
v_e	1	1	+1.001 157 53	1	1	-1.001 165 912	-0.5
v_μ	2	3	+0.978 240 60	1	1	-1.001 165 912	-0.5
v_τ	3	6	+0.863 605 77	1	1	-1.001 165 912	-0.5

Table 16-3 System g/2 factors

Particle	Predicted Mass Ratio (m/m_e)	Predicted Mass (m)	Measured [8] Mass (m) CL%	Range Mass (m)
ν_e	i0.000 001028	-i0.525 eV	<225 eV 95	<225-<460 eV
ν_μ	i 3.213	-i1.642 MeV	<0.19 MeV 90	<0.15-<0.65
ν_τ	i28.187	-i14.40 MeV	<18.2MeV 95	<1-<149

Table 16-4

We populate three more similar, but not identical, tables for anti-neutrinos.

Particle	B	N	$-B/N\alpha^{N/B}$	b	n	$b/n\alpha^{n/b}$	$(1-v_2^2/c^2)^{1/2}$
$\overline{\nu}_e$	1	1	-137.035 999	1	1	137.035 999	i0.007 071
$\overline{\nu}_\mu$	2	3	-1,069.450 64	1	1	137.035 999	i0.007 071
$\overline{\nu}_\tau$	3	6	-9,389.432 52	1	1	137.035 999	i0.007 071

Table 16-5 Anti-neutrino family parameters

In the next table, the g/2 factor to be utilized for B = 1 and N = 1 is the one for the anti-pion without the meso-electric factor. The g/2 factor for the second term is the anti-electron (positron) family value.

Particle	B_i	N_i	$g_i/2$	b_j	n_j	$g_j/2$	EF
$\overline{\nu}_e$	1	1	-1.001 157 533	0	0	+1.001 159 652	-0.5
$\overline{\nu}_\mu$	2	3	-1.001 165 912	0	0	+1.001 159 652	-0.5
$\overline{\nu}_\tau$	3	6	-1.001 157 653	0	0	+1.001 159 652	-0.5

Table 16-6 System g/2 factors

	Mass Ratio (m/m$_e$)	Mass (m)	Mass (m) CL%	Mass range (m)
$\bar{\nu}_e$	-i0.000 001 027	+i0.525 eV	<2 eV no data	<2 -<92 eV
$\bar{\nu}_\mu$	-i 3.300	+i1.686 MeV	no data	no data
$\bar{\nu}_\tau$	-i32.749	+i16.735 MeV	no data	no data

Table 16-7

The $\bar{\nu}_\mu$ and $\bar{\nu}_\tau$ have no measured limits in reference [8]. The $\bar{\nu}_e$ value is within the measured limits.

Based on our experience in calculating masses from first principles in the Electrino Fusion Model of Elementary Particles, we make two predictions regarding neutrinos and anti-neutrinos: The mass of the electron neutrino should be -i0.525 eV; and the mass of the electron anti-neutrino should be +i0.525 eV. The rest of the values are in the off chance that the v_2 for muon and tauon neutrinos and anti-neutrinos is also 1.000025 c.

[1]   http://www.ps.uci.edu/~superk/

[2]   http://en.wikipedia.org/wiki/Super-Kamiokande

[3]   http://arxiv.org/abs/0802.1041

[4]   Google: Robert Evans, "Particles found to break speed of light" (Reuters):

[5]   http://www.anti-relativity.com/

[6]   Gordon L. Ziegler, *Electrino Physics* (Printed in the United States of America, Xliebris LLC, November 3, 2011), Chapter 7.

[7]   Gordon L. Ziegler and Iris Irene Koch, *Advanced Electrino Physics*, Draft 2 (Printed in the United States of America, Xliebris LLC, 05/01/2014), Chapters 4 and 5.

[8]   J. Beringer *et al* (Particle Data Group) PR **D86**, 010001 (2012) and 2013 partial update for the 2014 edition (URL: http://pdg.lbl.gov).

Chapter 17

Gravitons

There is no formulation of gravitons in the Standard Model of Physics, though there is expectation that there should be such things as gravitons, and they should be spin 2. As a matter of fact, it is impossible to construct gravitons from quarks, though it is possible to construct several different kinds of gravitons from electrinos with the Electrino Hypothesis of the Electrino Fusion Model of Elementary Particles. In Chapter 8 there is a section on the structure of gravitons at the end of the chapter. In Chapter 7 are many Decay Modes solved by Chonomic reactions combining various gravitons with various parent particles, re-dividing into various daughter particles. In Chapter 7 we get a good feel for the different structures of gravitons and how they gravitationally combine with particles.

The authors of this set of volumes have had theoretical structures of gravitons for quite a few years now. The purpose of this chapter is to calculate from first principles the masses of the various kinds of gravitons using the roadmap for such calculations in chapters 9, 10, 11, 14 and 15. This should actually be simpler than the calculation of the masses of the neutrinos and anti-neutrinos in the last chapter. The big surprise in this chapter is that the elusive invisible gravitons do not have insignificant masses. Since Providence has ruled that all particles in the Universe, except the electron, should be calculable from first principles without definition, the echons in the chonomic grids for gravitons should be forced up to significant mass states. Gravitons, we will find out, should have real masses comparable to the masses of mesons.

The first graviton structure we wish to solve for from Chapter 8 is:

$g^{\pm}$

$$I(J) = 0(2)$$

Mass m not measured yet.

	g^+			g^-	
About 70.0 MeV.			About 70.0 MeV.		

```
  2      |              2      |
  1    + | +            1    - | -
  0      |              0      |


     1 | 2                -1 | -2
       |                     |
```

Higher energy state orbiting particle pairs can also compose $g^{\pm}$ gravitons.

It is easy to see that this kind of graviton is composed of an electron orbiting about a positron with the meso-electric force holding them together—opposites attracting. Unlike positive particles at higher energy states, the positron here has 0.000 000 000 for the meso-electric term, making the positron the simple charge conjugate of the electron. The problem is a two body problem (the balancing equation below must begin with a ½). We are dealing with whole particles orbiting here (the m in the balancing equation must be divided by 1). The electron and positron are whole opposite charges, so they are $(-e/1)$ and $(+e/1)$ in the balancing equation below.

$$\tfrac{1}{2}\,(m/1)\,v_0^2/r = (-e/1)(+e/1)\,1/4\pi\varepsilon_0\alpha^{N/B\,+1}\,r^2. \tag{17-1}$$

We can cancel an r out of both sides of the equation and factor out an

$$\hbar c = e^2/4\pi\varepsilon_0\alpha. \tag{17-2}$$

Using the techniques under Eq. (10-36), the force balancing equation (the meso-electric force and inertia) is reduced to the following equation:

$$mv_0^2 = -\hbar c/\alpha^{N/B} r \ . \tag{17-3}$$

We need now a spin relation for the graviton, and we have none. We do not have measured graviton masses to go by in this case. But it is not essential that we have the precise values for the masses of

223

the gravitons. We need only a ball park figure for the mass of each graviton. We have here a problem of whole particles orbiting. We have two precedents for whole particles (electrons) orbiting—the neutron and the neutrino—both having the following spin relation and no contrary spin relations. The assumed spin relation, then, taken as a postulate is

$$r = N\hbar / Bmc .$$
(9-5)

Combining Eqn. (17-3) with Eq. (9-5), we obtain

$$mv_o^2 = -\hbar cBmc/N\hbar\alpha^{N/B}$$
(17-4)

$$v_{og\pm}^2 = -B/N\alpha^{N/B} c^2$$
(17-5)

The mass of the electron in the graviton is 0.510998928 MeV. The mass of the positron orbiting is -0.510998928 MeV—the intrinsic masses of the electron and positron in the graviton cancel out. But the orbital mass does not cancel out. Providence has ruled that all particle masses in the Universe except electrons should be calculable from first principles without definition. To calculate the exponential polynomial for the $g^\pm$ graviton, that means the B and N lowest level of 0 and 0 must be ruled out in this case. The lowest possible values of the B and N in the exponential polynomial $g^\pm$ graviton case are 1 and 1. The exponential polynomial for this case then is $1/\alpha$ = 137.035 999. The EF (Energy Factor) is 1.0 (there is kinetic energy and there is potential energy—but not added; they are one and the same). The g/2 factor is not much, and can be neglected in this case. The m/m_e is then about 137, and the lowest real mass of the $g^\pm$ is about 70.0 MeV. Higher state gravitons would be convenient in some reactions (but not necessary) and could easily be solved for by using higher state B and N values of 2 and 3 and 3 and 6.

The next graviton we wish to solve for is an electron neutrino orbiting a positron anti-neutrino:

I(J) = 0(2)
Not yet measured

```
            g^o+
        About 70.0 MeV
    2       |
    1     +o|o+
    0       |


        -3|-2
          |
```

A negative + echon orbits a positive o echon with -1 orbital spin, forming a neutrino. A positive + echon orbits a negative o echon with -1 orbital spin, forming an anti-neutrino. The neutrino and anti- neutrino orbit each other with an additional -1 orbital spin. The result is a g^{o+} graviton with -2 spin.

.g^{o-}

I(J) = 0(2)   The spin conjugate graviton g^{o-} is also possible.
The mass is not measured yet

```
            g^o-
        About 70.0 MeV
    2       |
    1     -o|o-
    0       |


        3|2
         |
```

It is easy to see that this kind of graviton is composed of an electron neutrino orbiting about a positron anti-neutrino. With what force it does so is not as easy of a question! If the orbiting electrons and positrons cancel out the charges of the pions and anti-pions in this case, then the meso-electric force between the pions and anti-pions is zero, and the neutrinos must be held together by the strong gravitational force. But if the electrons and positrons orbited in-sink with the pions and anti-pions, and the pions were closest to the anti-pions in this problem, and the electrons and positrons were polarized to the periphery of the orbits, then the pion and anti-pion could be attracted to each other by the usual meso-electric force. Do these two pairs of particles polarize each other or not? It is easiest to assume that they do polarize each other, and the problem is not two neutrals attracting each other by strong force, but two opposite charged pions attracting each other by the meso-electric force—

which balancing equation is easy to write and is identical to Eqn. (16-1). Using the usual techniques, this reduces to Eqn. (16-3). Using the graviton spin equation this solves for a m/m_e of 137.035 999 and a mass of 70.0 MeV, notwithstanding that the neutrinos and anti-neutrinos have minus or plus imaginery masses.

The next and last graviton we want to solve the masses for is the g^o. According to chonomic decay schemes, a 1 spin graviton also exists.

g^o $I(J) = 0(1)$
         Mass not measured yet.
                 g^o
         About 70.0 MeV
    2 |
    1   o | o
    0 |

       1 | 1
         |

Like the other two types of gravitons, this type of graviton has the same balancing Eqn. as (17-1). The minimum gravitational mass is also about 70.0 MeV. That is not all! The higher mass states of all three types of gravitons will also be identical arrays of values. Higher state gravitons would have been convenient in some of the Decay Modes in Chapter 7, but not necessary.

The big surprise in this chapter is that the elusive invisible gravitons do not have insignificant masses! They all have masses in lowest state comparable to some lower state mesons—about 70.0 MeV. This is an aether model of physics. The authors find that the aether is composed of a large number of integral spin bosons— mostly 2 or 1 spin gravitons, but also some 0 spin particles—like π^0 particles.

Galaxies in this theory would have large clouds of invisible massive bosons—a source of calculated Dark Matter or Dark energy. The 70.0 Mev mass per graviton is the minimum value. Higher state gravitons are not calculated here by the authors. That is a simple exercise with a calculator left to the readers.

Chapter 18

Magnetons

Like gravitons, all magnitons are massive particles—either mesons or baryons. They have an echon pair of either $-+$ or $+-$ in the lowest non-definition state of 1 and 1 for the B and N values. In the $--$ or $++$ states, as in gravitons, the echons have opposite magnetic fields which cancel out. In the magnetons one echon is tipped upside down, making the particle more massive, but also making the magnetic fields of the sub-particles not cancel out, but add to a magnetic dipole, which can align with other ambient magnetic dipoles to make magnetic lines of force. Depending on the density of the magnetic dipoles in the aether, there is a limit to how strong a magnetic force can be achieved. The following is the lowest mass magneton:

$\omega(782)$ $I^G(J^{PC}) = 0^-(1^{--})$

Mass m = 781.94 ± 0.12 MeV   (S = 1.5)

$\omega(782)$

```
          ?
2         |
1       - | +
0         |

        1 | 1
          |
```

More magnetons are $\omega(1420)$, $\omega(1650)$, ω_3 (1670), $K_1(1270)$, J/ψ (1S), ψ (2S), ψ (3770), χ (3872), Y(10860), N (1535)½⁻, Λ (1810)½⁺, Ω^-, Ω (2250)⁻, and Ω (2770)⁰. We have the measured masses of all those particles already (see Chapter 8). Soon (like maybe in a six months) we will have the calculated and predicted masses of each particle from first principles also in *Predicting the Masses*, Volume 2, Predicting the Mesons and *Predicting the Masses*, Volume 3, Predicting the Baryons. They won't be much different than the measured values.

228

www.ingramcontent.com/pod-product-compliance
Lightning Source LLC
Chambersburg PA
CBHW051902170526
45168CB00001B/208